Modern Technologies of Thin Films Deposition
Chemical Phosphatation

Andrei Victor SANDU
Costică BEJINARIU
Ioan Gabriel SANDU
Mohd Mustafa Al Bakri ABDULLAH

Published by **Materials Research Forum LLC**
Millersville, PA 17551, USA

Published as part of the book series
Materials Research Foundations
Volume 39 (2018)
ISSN 2471-8890 (Print)
ISSN 2471-8904 (Online)

Print ISBN 978-1-945291-90-6
ePDF ISBN 978-1-945291-91-3

Distributed worldwide by

Materials Research Forum LLC
105 Springdale Lane
Millersville, PA 17551
USA
http://www.mrforum.com

Printed in the United States of America
10 9 8 7 6 5 4 3 2 1

Table of Contents

Introduction

Thin phosphate layers, used as precast processing and during installation of parts, are very attractive technologically and in terms of reliability, the manufacturing cost and environmental impact.

Research in this area has been driven by use of simple co-precipitation processes in acidic aqueous solutions with minimum consumption of energy and materials.

This book is based on research conducted in a PhD thesis and is part of the efforts to solve scientific and technological problems related to obtaining thin layers by phosphating, with corrosion protection and lubricating role, both in plastic deformation processing and protection as finished parts.

The research had as main purpose the development of phosphating technologies with high reliability, minimal use of materials and energy and with minimal impact on the environment and operator.

The purpose of this volume is to present alternatives to obtain and characterize phosphate layers by the selection of co-precipitation cation and surfactants, allowing the production of dendritic structures with optimal chemical, physical-structural and mechanical properties.

The research conducted in the thesis was conducted in a complex interdisciplinary system, within recognized collective in materials science and engineering from three renowned universities in Europe, respectively: Gheorghe Asachi Technical University of Iasi, Alexandru Ioan Cuza University of Iasi and Universite de Technologie Belfort-Montbeliard, France.

In research, we used modern methods and techniques of scientific investigation, some acquired during research, which allowed conducting training sessions with specialists manufacturing companies and others with experts and academics from the respective laboratories.

In the development of phosphating technologies we followed the addition of, on the one hand with a series of metal cations and, on the other hand, with surface active agents, with roles in the processes of nucleation, allowing to obtain compact layers, uniform, adherent to the substrate and with multiple practical implications.

Authors

Chapter 1

Known Phosphating Technologies

The corrosion protection of steel is a problem of high economic importance. The most used steel sheets in the industry have very low resistance to corrosion and can be easily penetrated by rust.

Here are the usual metal-coating procedures: hot-dip galvanizing (HDG), galvanneal (GA), electrogalvanizing (EG), electroplated zinc-nickel coating (ZnNi) and zinc-iron alloy coating (ZnFe). The hotdip process (HDG) is still widely used today, the formed layer being 6 to 20 micrometers thick. Galvanneal (GA) is a hot-dip zinc-coated sheet which has been heated after the application of the coating in order to allow interdiffusion of zinc and iron, being 6 to 11 micrometers thick. Electroplated zinc-coated sheet (EG) exhibit a very uniform coating thickness. Electroplated zinc-nickel alloy coatings (ZnNi) generally contain 10 to 14% nickel with a thickness of 20-40 micrometers.

Regarding the subsequent phosphating process, it starts with a degreasing step in mild alkaline baths followed by pickling in an acid solution and immersion in the phosphating bath. The chemical composition and the crystallographic structure of this layer depend on the compositions of the substrate and that of the phosphating bath. Two of the common phosphate structures are $Zn_3(PO_4)_2 \cdot 4H_2O$ (hopeite) and $Zn_2Fe(PO_4)_2 \cdot 4H_2O$ (phosphophylite).

Zn phosphating is one of the most widely used surface treatments and is generally conducted in acid aqueous phosphate solutions containing zinc ions and phosphoric acid, as well as nitrate ions as accelerator to promote the oxidation and dissolution of the metallic surface. It is known that the phosphate coating develops via the nucleation, growth, and coalescence of zinc phosphate grains. The corrosion resistance of the phosphate coating is related to the size and population density of pores in the coating; that is, the pores provide a path for corrosion attack.

There are known solutions with additives, such as Ni^{2+}, Mn^{2+}, and Mg^{2+}, helping to form a homogeneous phosphate coating with finer zinc phosphate grains. The presence of Ni^{2+} and Mn^{2+} in the phosphate solution apparently affects the nucleation and growth of the zinc phosphate grains.

The crystalline phosphatation processes take place after an electrochemical mechanism with „common electrode", in which the anodic processes are of the polyelectrodic type

1

(micropiles in short-circuit) and cause an interaction between the low soluble layer of pyrophosphate and the iron substrate. The Fe^{2+} ions formed in the anodic area (dissolving of the metal) contribute to the formation of the primary layer of zinc pyrophosphate, the resulted crystalline structures becoming inert to subsequent oxidative processes. In the cathodic area, hydrogen is liberated. The two electrochemical processes can take place simultaneously or not, resulting in small polarization areas with electron transfers through the iron substrate. Generally, the phosphatation process takes place in one step.

Literature describes many anticorrosion procedures of iron object surfaces by chemical phosphatation, which consist of the precipitation of thin, continuous and uniform layer of low soluble pyrophosphates of zinc, nickel, manganese or calcium and iron. The most used metal is zinc. It can be used in mixtures or alone, depending on the purpose of the layer.

It is well known that the coating layers deposited by conversion on metal surfaces, which involves chemical precipitation processes with the formation of poorly soluble and strongly substrate-adherent compounds, have physical, chemical and mechanical properties dependent on the architecture of surfaces, on the nature of the covering technology parameters and materials [*Oniciu, 1980; Marinescu, 1984; Grünwald, 1995; Schlesinger, 2000*].

Among these, the phosphating layers have been the most frequently studied and used, due to the fact that these pellicles have special characteristics: corrosion protection, painting and dipping substrate, solid lubricant in the cold working process of metals, esthetic role (polychromies), etc.

Corrosion protection by phosphating is due to the formation of a new, even, compact and passivator layer adhering to the substrate, which consists of a mixture of secondary and tertiary phosphates of the metal cations in the four blocks of poorly soluble chemical elements ("d", "p", "f" and "s"), obtained by sequential coprecipitation or precipitation processes comprising one or several stages, after the prior degreasing and pickling of iron-based metal surfaces [*Rausch, 1990; Grünwald, 1995; Schlesinger, 2000; Marcus, 2002*].

In order to achieve special corrosion protection properties, zinc phosphate coatings and other cations of transition metals or of "p" and "s" block metals are used, which are subsequently dipped in oils or different film-forming passivators [*Oniciu, 1980; Marinescu, 1984; Rausch, 1990; Crow, 1994*].

Zinc phosphate solutions and cations of other metals susceptible of forming polychromies, like for instance Mo^{3+}, Cr^{3+}, Mn^{2+}, Co^{3+}, Y^{3+}, Gd^{3+}, Sb^{3+}, Bi^{3+}, Al^{3+}, Ce^{2+},

Mg^{2+}, Ca^{2+} etc, may be used to achieve a beautiful, shiny and evenly crystallized structure, with monochromatic or hued ornamental drawing (texture). These films may be final or provisional and they are usually obtained on electroplated zinc, aluminum, etc. coatings, which increase the manufacturing costs considerably [*Sandu, 2002 and 2009; Ishizuka, 2003; Li, 2004; Narayanasamy, 2005*].

Different technologies are known which are employed to achieve thin zinc phosphate and nickel/cobalt/manganese coatings, both by chemical precipitation, and by electrochemical processes of cathodic coating or anodic oxidation, followed or not by heat treatments, with added fluorides, which also have corrosion protection properties.

When the purpose is the creation of a substrate for later painting, tertiary zinc phosphate is used, with high intra-crystalline porosity. The advantage of this procedure is the fact that zinc phosphate is dielectric and allows electrophoresis painting [*Oniciu, 1980; Marinescu, 1984; Grünwald, 1995*].

Phosphate-coated parts are often used in the car industry, due to their lubricating characteristics, since they reduce considerably the force necessary for cold and even semi-hot plastic deformation. In this respect, they are dipped in mineral or vegetable oils, molybdenum disulphide, colloidal graphite, etc. Thus, coatings with complex structure are obtained, based on a double decomposition reaction, with the formation of zinc soaps and tertiary sodium, potassium, etc. phosphates, according to the reaction:

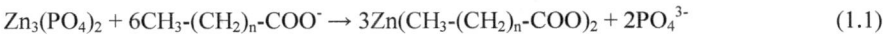

$$Zn_3(PO_4)_2 + 6CH_3\text{-}(CH_2)_n\text{-}COO^- \rightarrow 3Zn(CH_3\text{-}(CH_2)_n\text{-}COO)_2 + 2PO_4^{3-} \qquad (1.1)$$

where n is an even number between 14 and 18.

It was noticed that this reaction supports both cold deformation and tool wear reduction.

These properties are due to the fact that the temperature at the surface of the part is as high as 200°C, as a result of the very high mechanical stress when the phosphate in the passivator film undergoes a dehydration process (for instance, it loses two out of four molecules of crystal-hydrate), and cleavage planes occur which facilitate crystalline particle transfer during the deformation [*Rausch, 1990; Grünwald, 1995; Schlesinger, 2000*].

In order to reduce the friction coefficient and prevent jamming, primary manganese and nickel phosphate coatings are used, which are dipped in solid or liquid lubricants.

Ferrous ions are used to accelerate the oxidation processes, and a $(Mn^{II}/Fe^{II})_5H_2(PO_4)_4 \cdot 4H_2O$ coating with or without nickel occurs on a ferrous oxide and phosphate substrate.

In addition to crystalline phosphating, amorphous layers of iron phosphates are achieved. The parts thus treated may be used for light deformations, in poorly corrosive atmospheres or for painting purposes.

In industry, phosphating is applied both on ferrous parts (cast iron and steel), and on non-ferrous parts (aluminum, magnesium and zinc).

According to the international current trends, parts used in the car industry (valves, sparking plug bodies, union nuts and other special nuts, lubricators, nozzles, caps, regular nuts and locknuts, etc.) are manufactured by volumic plastic working. This requires the use of a phosphating technology, which allows the creation of highly porous thin passivator coatings that contain a large amount of lubricant and prevent streaks on the surface of processed parts [*Yoshio, 1988; Berki, 1998; Urata, 2000 and 2002; Kazuya, 2000; Bejinariu, 2011*].

1.1 Specific Phosphating Processes

An example of phosphating solution includes orthophosphoric acid, primary metal phosphate and an accelerator. Usually, other acids, such as nitric, fluorosilicic, fluorhydric or boric acids, are also used to accelerate the reactions [*Oniciu, 1980; Marinescu, 1984; Rausch, 1990; Grünwald, 1995*].

The most frequently used divalent metal is zinc, followed by manganese, nickel, calcium and very rarely iron. They be used independently or in different combinations.

Sodium or ammonium chromates or molybdates are used as reaction moderators, and sodium salts of aryl- or alkyl-sulfonate organic acids are used as active surface or wetting agents [*Rausch, 1990; Grünwald, 1995*].

The optimum pH of immersion solutions is 1.8 ... 3.2, whereas the optimum pH for spraying solutions is 2.5...3.2 [*Marinescu, 1984; Rausch, 1990; Grünwald, 1995*].

The solutions may be used at room temperature or heated up to 100°C [*Oniciu, 1980; Marinescu, 1984; Rausch, 1990*].

In chemical phosphating processes, metal surfaces act as polyelectrodes and require the existence of short-circuited micropiles, which include anodic areas where the metal is dissolved and cathodic areas where hydrogen is released. These two reactions may occur simultaneously or successively, with the occurrence of discrete polarization areas with electron transfer [*Rausch, 1990; Crow, 1994; Grünwald, 1995; Marcus, 2002*].

All phosphating processes start with a reaction of this type, the equilibrium of which may be favorably displaced by the use of oxidants (for example, NO_3^-, $Cr_2O_7^{2-}$, MoO_4^{2-}, WO_4^{2-} etc. anions), which act as electron and proton consumers.

The process is depicted in figure 1.1, being adapted from several representations [*Marinescu, 1984; Grünwald, 1995*].

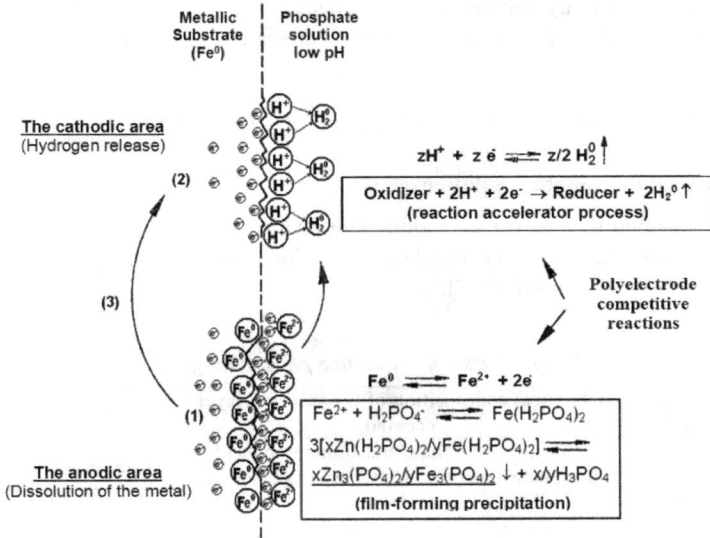

Fig. 1.1. Schematic representation of the chemical phosphating process:
1 – anodic area; 2 – cathodic area; 3 – electron transfer in metal.

The progress of the process in the sense of poorly soluble tertiary phosphate film formation depends on the existence of well-defined ratios between the precipitation compounds formed, various ions and other molecular species in the solution, which by competitive reactions determine free phosphoric acid consumption or formation.

The reaction specific to tertiary phosphate film formation out of primary phosphate firm formation is:

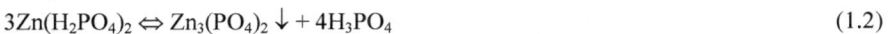

$$3Zn(H_2PO_4)_2 \Leftrightarrow Zn_3(PO_4)_2 \downarrow + 4H_3PO_4 \qquad (1.2)$$

Similarly, in the anodic area, the Fe^{2+} ion initially reacts with the dihydrophosphate anion and forms the primary salt:

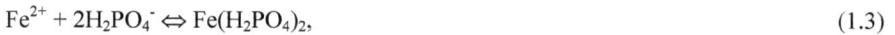

$$Fe^{2+} + 2H_2PO_4^- \Leftrightarrow Fe(H_2PO_4)_2, \tag{1.3}$$

and the excessive Fe^{2+} ions displace the equilibrium of the reaction of conversion of the primary salt into tertiary salt and cause the occurrence of Zn(II) and Fe(II) ortho and pyrophosphates, where the *phosphophyllite* predominates ($Zn_2Fe(PO_4)_2 \cdot 4H_2O$) [*Sandu, 2012*]:

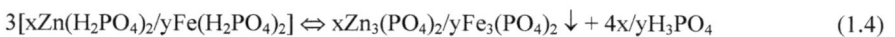

$$3[xZn(H_2PO_4)_2/yFe(H_2PO_4)_2] \Leftrightarrow xZn_3(PO_4)_2/yFe_3(PO_4)_2 \downarrow + 4x/yH_3PO_4 \tag{1.4}$$

1.2   Classical Phosphating Solutions

The determination of complex crystalline phosphating compositions specific to certain scopes has led to the development of defined solutions, shown in table 1.1 [*Oniciu, 1980; Marinescu, 1984; Grünwald, 1995*].

Table 1.1. Classical crystalline phosphating solutions.

Phosphating solution	Chemical composition of the solution/layer created	Phosphating scope
I	Zinc acid phosphate and ammonium acid phosphate + nitric acid /tertiary iron and zinc phosphates	Cold plastic working Corrosion protection
III	Manganese acid phosphate / manganese and iron phosphate	Extrusion
IV	Zinc acid phosphate + HNO_3/tertiary zinc phosphate	Painting substrate
VI	Zinc acid phosphate /tertiary zinc phosphate	Lubrication substrate
VII	Primary zinc phosphate + HNO_3/ tertiary zinc phosphate	Wire drawing
VIII A – VIII E	Primary zinc phosphate + HNO_3, $Ni(NO_3)_2$ and $HSiF_3$/tertiary zinc phosphate	Corrosion protection Painting substrate
IX	Primary manganese phosphate + HNO_3, $Ni(NO_3)_2$/ manganese and iron phosphate	Lubricant carrier
X A – X E	Chromic acid, free and complex fluorides / aluminum oxide layer	Corrosion protection Painting substrate
XI A – XI E	Primary manganese phosphate + HNO_3 and $Ni(NO_3)_2$/ tertiary manganese, nickel and iron phosphates	Lubricant carrier
XII A – XII E	Zinc phosphate + HNO_3 and $Ni(NO_3)_2$/ Tertiary zinc phosphates	Corrosion protection Painting substrate

As mentioned earlier, the coatings achieved may be crystalline and amorphous. Thus, table 1.2 also shows several solutions for the creation of amorphous thin coatings [*Marinescu, 1984; Yoshio, 1988; Rausch, 1990; Grünwald, 1995; Schlesinger, 2000; Marcus, 2002*].

Table 1.2. *Classical amorphous phosphating solutions.*

Trade name	Solution composition/created coatings	Phosphating scope
Degreasing agent-Passivator I	Alkaline phosphates + organic surfactant/$FePO_4$ + Fe_2O_3	Painting substrate
Degreasing agent-Passivator II	Alkaline phosphates + chlorate + molybdate + organic surfactant/$FePO_4$ + Fe_2O_3	Painting substrate
Pickling-phosphating salt	Phosphoric acid + organic surface-active agent /$FePO_4$	Painting substrate Preliminary treatment

The amorphous phosphating process is conducted by using salt solutions of different types, such as alkaline ammonium phosphates and/or organic substances. In this case, a substitution reaction occurs instead of precipitation, where the cations in the solution do not occur in the coating; the protection coating is usually made up of ferric phosphate and iron oxide. Thus, the basic metal contributes to corrosion coating forming by itself and has the same role and behavior as the coating created by anodic oxidation. In these procedures, the pH of the solutions used does not exceed 6.0 [*Askienazy, 1980; Yoshio, 1988; Gosset, 1989, Song-Gu, 1988; Dong-Wook, 1990; Harry, 1991; Jae-Ryung, 1992; Sigeki, 1992; Taranu, 1997; Tamotsu, 1999*].

1.3 Modern Phosphating Procedures

This subsection describes classical phosphating solutions and modern processes but also the study of the morphology and structure of thin films based on the obtaining processes showing phosphating with pre-treatment and the role of other metallic cations (Ni, Nb, Co, Mn, Mg) and the role of accelerators and other surface-active agents, followed by zinc phosphates with specific color and the role of post-treatments.

A large number of phosphating solutions used on metallic surfaces, which contain Fe, Al, Zn etc., are also described in literature. Among these, we will only tackle hereunder the ones that have already been patented and published, which allow the creation of thin coatings, mainly inorganic systems. Here are some of them:

- composition for cold phosphating of metallic surfaces, which are used as corrosion protection agents and prepare the surfaces for painting or other covers. The solution is made up of: 1.7 density phosphoric acid, 210-240g/L, 45-60 g/L zinc oxide, 8-12g/L sodium nitrate and 1.2-2.4 chrome trioxide [*Trusov, 2004*];

- also for corrosion protection purposes, *N.V. Varentsova and V.A. Chumaevskij* (1998) created a phosphating solution based on 0.03-0.5 g/L nickel ions, (PO_4^{3-}) 1.7-7g/L nickel ions, (NO_3^-) 1.5-5.2 nitrate ions, 0.03-0.5 fluoride ions, 3-25g/L zinc-based by-products, which contains 25% Zn, min. 25% P_2O_5, 4-8% manganese and water, in addition to by-products based on 2-20 g/L manganese nitrate, which contains 80% Mn(II), aluminum perchlorate, formic acid, manganese citrate and manganese hydroxide. The role of this coating is to reduce friction in the plastic deformation processes and to increase corrosion protection;

- another acid phosphating solution containing at least two metallic ions, one of which may be Zn, Mn or Fe. It includes a polyphosphate with the formula $(MPO_3)_n$ where n $\geq$ 3 and M is an alkaline, alkaline-earthy element or ammonium and a chelation agent [*Gosset, 1989, Tamotsu, 1999*];

- a treatment applied by spraying or immersion on steel and cast iron surfaces, which improves corrosion protection and may be used as painting substrate, is based on solutions containing tungstate or molybdate ions with pH ranging from 5.8 to 6.5 [*Askienazy, 1980*].

Another phosphating solution was described by *I. Ţăranu et al.* (1997), which may contain 2 to 50% phosphoric acid, 10 to 90% ethanol and water. Here are some examples: 2% phosphoric acid, 10% ethanol and 88% water; 50% phosphoric acid, 10 % ethanol and for the rest water; 2% phosphoric acid, 90% ethanol and 8% water.

M. Schoenherr et al.'s invention (2009) relies on a zirconium or titan fluoro-complex and phosphate ions. This method is suitable for the pre-treatment of paints applied by electrophoresis, for instance on radiators.

Z. Zhang (2009) developed a microcrystalline phosphating solution for brake components which contains 40...50 parts zinc nitrate, 22...30 parts zinc dihydrogen phosphate, **30...40 parts Mazhev salt** $(Mn(H_2PO_4)_2 \cdot 2H_2O)$, 1.5...2 parts guanidine nitrate, 1.5...2 parts sodium m-nitrobenzene sulfonate, 1...2 parts sodium tripolyphosphate, 2...4 parts tartaric acid, 2...3 parts sulfosalicylic acid, 1.5...2.5 parts citric acid and 1000...1200 parts water. This solution is used for iron part phosphating treatment, which is a fast process with good performance and low environmental impact. The crystals formed are finely grained, have even colors and good adherence.

A. Paukson (2011) suggested a phosphating procedure based on 100 gravimetric parts of orthophosphoric acid (85%), 10…19 parts Fe(III) oxide, 1.1…1.6 parts Al, Zn and Mg metallic powders, 10…30 parts water, 1…3 parts calcium and 0.5…1.5 parts citric acid, which is used in industrial and civil engineering, agriculture and wood processing, as it has a fireproof, esthetic and antiseptic role, and it supports acrylic or water paints.

D. Elbick et al. (2011) are the creators of a patented hybrid procedure designed for metallic surface treatment with a very varied and complex composition (Cr, Cu, Mn, Mo, Ag, Au, Pt, Pd, Rh, Pb, Sn, Ni, Zn and Fe) which, in addition to chemical processes, uses anodic oxidation with specific solutions containing oxalate, tungsten, molybdate, silicate, borate, nitrate, phosphite and phosphate anions as such or in combinations of two or more components. The solution pH should not exceed 6.0, and the oxidation potential should range between 0.5 and 20.0V, for a current density ranging from 1.0 to 200.0 A/cm^2. After oxidation, the surface may be polymer coated.

For steels with complex compositions containing Si, Ti, Zr, Hf, V, Ta, Ca, Ce, Sc, Nb, Cr, W, Al, Sr, Zn, Mg and Mo, *T. Konishi et al.* (2011) also used an aqueous solution based on nitrites and phosphate ions, which, by condensation in the presence of the ions fluoride, carbonate and silicate, for a pH ranging from 4.0 to 14.0, forms continuous oxidation-resistant films.

One of *J. Day's* inventions (2011) refers to a microporous phosphating solution containing 100...120 parts $Zn(NO_2)_2 \cdot 6H_2O$, 25...35 parts manganese dihydrogen phosphate, 2…6 parts sodium nitrate, 0.5…3.6 parts acceleration agent and mineral acids up to 80 parts, used in the automotive industry, aviation industry and arms industry. The coating achieved allows the inclusion of lubricating oils.

For the creation of thin films of electronically conducting metal compounds, *K. Okada et al.* (2011) suggest a technology consisting of the initial coating of the basic Fe part by a Zn-Ni alloy, which does not contain chromium, which is then turned into polyurethane resins by means of a solution based on phosphates, vanadates, resins which are deposited in a 10 to 1000mg/m^2 thin and even layer.

G. Qiu et al (2010) developed a crystalline phosphating procedure using aqueous solutions based on manganese phosphate which is formed by precipitation in the presence of dihydrogen phosphate, orthophosphoric acid and sodium hypochlorite as oxidant, at a temperature of 40…100 °C, for 1 hour to 7 days $(Mn_5(HPO_4)_2(PO_4)_2 \cdot H_2O)$, as such or as hureolite $(Mn_{5-x}M_x(HPO_4)_2(PO_4)_2 \cdot H_2O)$, obtained by conversion, where M may be Co,

Fe, Ni, Cu or Zn. In addition to their protective role, such thin films are used in catalytic processes, optoelectronic processes and magnetic passive components.

A chromium-free phosphating solution is known, in normal temperature conditions and lasting only a few minutes (6…10), for parts based on Mg-Zn alloys, which contains 110…130 g K_2HPO_4 12…18 g $KMnO_4$ and 6…10 g of citric acid in a liter of water. The solution pH is adjusted by means of H_3PO_4 within the 4.5…6.5 range. The procedure has the advantage of obtaining an inexpensive, even, thin and substrate-adherent coating, which is very resistant to corrosion and has a minimum impact on the environment [*Guan, 2010*].

H. Matsuda et al. (2010) also carried out a metal part phosphating procedure, in which the thin protection film is chromium free. The solution they used contains vanadium pentoxide, calcium vanadate and/or ammonium metavanadate, in addition to Zn, Al or Mg phosphate dispersed in aqueous sodium chloride solution as fine nanodispersions of ceramic pigments, which are deposited by coprecipitation on metal surfaces.

D.E. Chasan and M. Ribeaud (2010) obtained a thin film of organic systems providing corrosion protection, which, in addition to a lubricant, also contains a set of corrosion inhibitors based on benzotriazole, alkyl borates and alkyl phosphates, in the presence of ammonium phosphate. Motor oil was used as lubricant, which contains 1-(di-isooctylaminomethyl)-1,2,4-triazole or 1-(di-(2-ethyllhexyl)aminomethyl)-1,2,4-triazole, in short benzotriazole.

Y. Xu et al. (2012) described a siliceous steel coating technology using Zn(II), Al(III), Ca(II) and Mg(II) dihydrogen phosphates ($Zn(H_2PO_4)_2$, $Al(H_2PO_4)_3$, $Ca(H_2PO_4)_2$ or $Mg(H_2PO_4)_2$), coated by 10-60 parts epoxy resin, 0.001-10 parts zinc naphthenate or isooctate, up to 100 parts organic solvent.

1.4 Study of the Morphology and Structure of Thin Films depending on the Production Process

1.4.1 Phosphating with pre-treatment

The research team [*Sandu, 2011*] carried out phosphate coating procedures in order to obtain polychromic and passivator films. After degreasing and pickling, we performed a two-stage treatment [*Bejinariu, 2010, 2011a and 2012b*]:

- first, a copper coating was deposited by subtraction, according to the reaction:

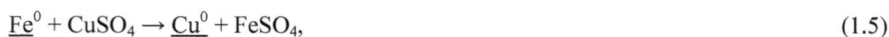

$$\underline{Fe}^0 + CuSO_4 \rightarrow \underline{Cu}^0 + FeSO_4, \tag{1.5}$$

- then the copper orthophosphate was precipitated by its immersion in two acid aqueous solutions, one having poor oxidizing characteristics, based on phosphate anions and zinc cations, in the presence of permanganate anions as surface moderator (reaction 1.6), and the other based on orthophosphoric acid containing dispersed metal zinc powder, by immersion at room temperature on clean surfaces, according to reaction 1.7:

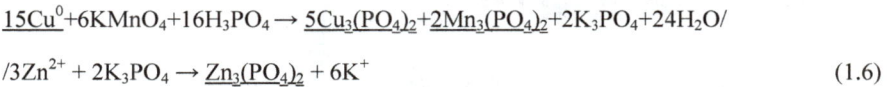

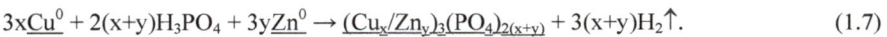

$$15\underline{Cu^0}+6KMnO_4+16H_3PO_4 \rightarrow \underline{5Cu_3(PO_4)_2}+\underline{2Mn_3(PO_4)_2}+2K_3PO_4+24H_2O/$$

$$/3Zn^{2+} + 2K_3PO_4 \rightarrow \underline{Zn_3(PO_4)_2} + 6K^+ \tag{1.6}$$

$$3x\underline{Cu^0} + 2(x+y)H_3PO_4 + 3y\underline{Zn^0} \rightarrow \underline{(Cu_x/Zn_y)_3(PO_4)_{2(x+y)}} + 3(x+y)H_2\uparrow. \tag{1.7}$$

This procedure [*Sandu, 2011*] is based on phosphating processes in acid environment by additive-subtractive mechanisms, which use Cu, Zn and Mn cations in various oxidation states, which become inert after precipitation towards the Fe(0) substrate. Phosphate ions react with the iron support and form a poorly soluble thin film, which is active towards the zinc and copper ion substitution processes

Figure 1.2 shows the microscopic image of the standard phosphate film, whereas figure 1.3 depicts the microscopic image of the phosphate film with the Cu film and also the EDX spectra pertaining to the samples.

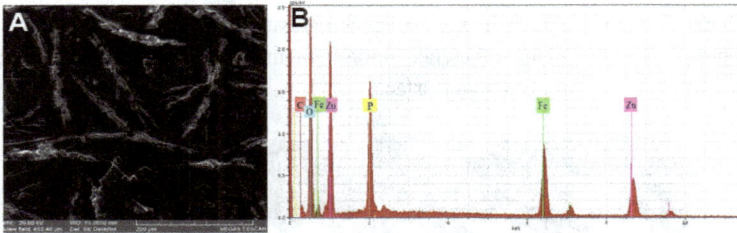

Fig. 1.2. *Microphotography of the phosphate-coated surface (A) and EDX spectrum (B).*

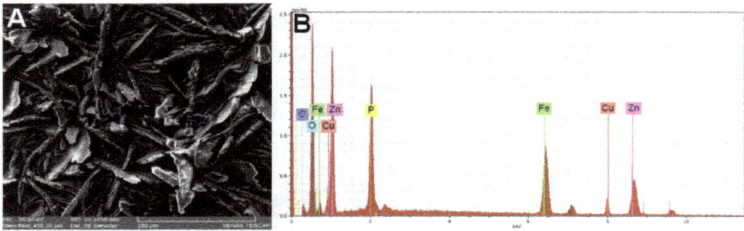

Fig. 1.3. *SEM image of the zinc phosphate with Cu (A) and EDX spectrum (B).*

H. Nikdehghan et al. (2008) studied the effect of substrate heat treatments before the coating. The iron substrate underwent 880 °C heat treatments. Then, the samples were immersed for 16 minutes in a 80 °C zinc-manganese phosphate solution with molybdenum ions added. The findings showed that the substrate microstructure has a considerable effect on Zn-Mn phosphate film morphology and that the film is influenced by the heat treatment applied before the coating. The coating on a hardened substrate exhibited the best corrosion resistance, whereas the corrosion resistance of hardened and tempered surfaces is superior to that of surfaces that have not undergone heat treatments.

1.4.2   The role of other metal cations in the phosphating baths

The influence of Ni^{2+} cations in the phosphating baths consists of corrosion stability improvement, and also layer adherence improvement [*Zimmerman, 2003*]. Figure 1.4 shows the SEM images of phosphate surfaces.

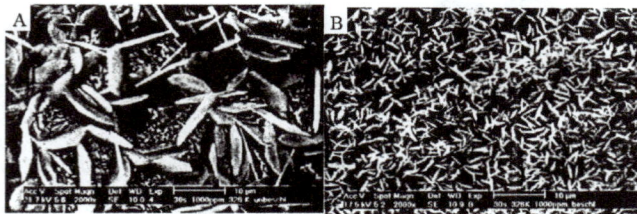

Fig. 1.4. *SEM images of the phosphate with 1000ppm Ni^{2+} surface:*
(A) without accelerator (sodium nitrate) after 30s and (B) with accelerator after 30s.

The Ni deposits begin in the corroded areas after the sample has been immersed in the phosphating bath. The corrosion protection characteristics are enhanced, being caused by the formation of nickel deposits at the foot of the pores in the phosphate layer. They not only catalyze surface reactions, but they also generate a much stronger surface.

In other researches [*Banczek, 2008*], nickel was replaced by niobium in zinc phosphate baths (ZnP) and phosphate deposits were proven to have increased. Also, the forming time was much shorter for niobium than for nickel. The XRD chemical analysis enabled them to conclude that the following compounds were formed: $Zn_3(PO_4)_2.4H_2O$ (hopeite) and $Zn_2Fe(PO_4)_2.4H_2O$ (phosphophyllite).

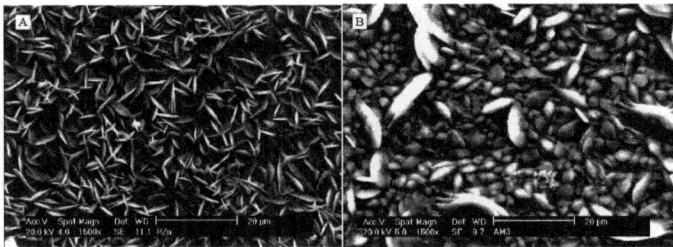

Fig. 1.5. *Micrographs of the carbon steel (SAE 1010) surface in (A) PZn+Ni bath (NaNO₂ 0.5 g/L;T = 25°C, t = 5 min), (B) PZn+Nb, (NaNO₂ 2 g/L, T = 25°C, t = 3 min).*

According to the SEM images (Fig. 1.5), the layers formed in ZnP +Ni baths consist of needle-shaped crystals, as compared to granular ones found in ZnP-Nb baths, which exhibit better surface coating and hence better resistance to corrosion. The authors concluded that niobium may replace nickel in phosphate baths and is more cost-effective [*Banczek, 2008; Girciene, 2003*].

A. Zarei and A. Afshar (2009) created a zinc phosphate solution with good covering characteristics at low temperatures, which involves cations like calcium and nickel. Stable phosphate samples with or without calcium and nickel salts were prepared, and their morphology and structure were determined by SEM and XRD. The findings proved that the incorporated calcium and nickel change the layer structure from phosphophyllite-hopeite to scholzite- phosphonicolite-phosphophyllite-hopeite with the considerable reduction of the size of the grains and improved compactness. They were proven by spray test with saline solutions.

Materials Research Forum LLC
doi: http://dx.doi.org/10.21741/9781945291913

Cobalt (Co^{2+}) and nickel (Ni^{2+}) ion incorporation into zinc phosphate solution altered by polyacrylic acid (p(AA)), which promotes crystalline zinc phosphate growth and development. These improvements are due first and foremost to the preferential combination of Co^{2+} and Ni^{2+} ions with p(AA), in order to form complex Co- and Ni-OOC- salts. These deposits provide better steel coating, which contribute to corrosion protection [*Sugama, 1992*].

An invention by *V.I. Trusov and V.L. Kiselev* (2004) relies on a composition used to create phosphate-coated surfaces with corrosion protection properties, as a preparation film for paints. This solution includes: phosphoric acid (ρ=1.7, 210-240 g/L), zinc oxide (45-60 g/L), sodium nitrate (8-12 g/L) and chromium trioxide (1.2-2.4 g/L). The film thus formed is a very good corrosion inhibitor.

Another coating with good corrosion resistance was achieved by *N.V. Varentsova and V.A. Chumaevskij* (1998) by using a solution of: nickel ions (0.03-0.5 g/L), phosphate ions (P_2O_5 1.7-7 g/L), nitrate ions (1.5-5.2 g/L), fluoride ions (0.03-0.5 g/L), zinc waste concentrates (3-25 g/L: 25% zinc, min. 25% P_2O_5, 4-8% manganese and water), manganese nitrate waste (2-20 g/L, Mn(II) 80%, aluminum perchlorate, formic acid, manganese citrate and hydroxide). The film thus created has special corrosion resistance [*Harry, 1991*].

Z. Zhang (2009) created a crystalline phosphate solution for brake blocks, which contained 40..50 parts zinc nitrate, 22…30 parts zinc dihydrogen phosphate, 30…40 parts Mazhev salt ($Mn(H_2PO_4)_2 \cdot 2H_2O$), 1.5…2 parts guanidine nitrate, 1.5…2 parts sodium m-nitrobenzene sulfonate, 1…2 parts sodium tripolyphosphate, 2…4 parts tartaric acid, 2…3 parts sulfosalicylic acid, 1.5…2.5 parts citric acid and 1000…1200 parts water. The process is quick and has low environmental impact. The formed crystals are even, small and have good adherence.

Tsai et al. (2010a, 2010b and 2011) developed a phosphate solution made up of 18.4 mM zinc oxide (ZnO), 235 mM sodium nitrate, 220 mM phosphoric acid and 0–117 mM magnesium nitrate at 45 °C. Mg^{2+} in phosphate solutions has the following impact: reduces the size of the grains and increases phosphate grain population density, the increased magnesium concentration increases resistance to corrosion, by reducing film porosity. (Fig. 1.6).

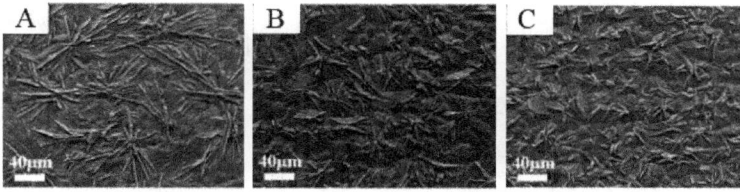

Fig. 1.6. *SEM image of the coating obtained in the phosphate solution containing (A) 39 mM Mg^{2+}, (B) 78 mM Mg^{2+}, and (C) 117 mM Mg^{2+}, for 120s. [33].*

I. Kiyokazu and S. Hidetosh (2003) suggested another zinc phosphate coating procedure containing at least 2% Mg and 0.01 to 1% at least one of the following elements: Ni, Co and Cu, thus creating a film of minimum 0.7 g/m^2.

1.4.3 The role of accelerators and other surface-active agents

The precipitation conditions mainly require the presence an oxidant accelerator, such as: dichromate, nitrate, chloride, persulfate, acid pH (1.8…3.2) and temperature ranging from 40 to 200 °C.

M. Sheng et al. (2011) added nano-SiO_2, which is used as accelerator designed to improve corrosion resistance of carbon steels. Nano-SiO_2 has been shown to have special influence on the morphology of films with 3 g/L nano-SiO_2 concentration, when the smallest roughness is noticed (Fig. 1.7). The electrochemical characterization showed that nano-SiO_2 containing phosphate has much better corrosion resistance than phosphate that does not contain it. Nano-SiO_2 may be used as additive, being less polluting than other agents.

Fig. 1.7. *SEM image of phosphate films containing nano-SiO_2: (A) 2 g/L, (B) 3 g/L.*

Materials Research Forum LLC
doi: http://dx.doi.org/10.21741/9781945291913

Benzotriazole (BTAH) is also an additive determining a more compact phosphate film [*Banczek, 2006*]. The films achieved have low corrosion resistance in an acid environment, but good resistance in saline environment. BTAH supports the nucleation processes which result in small-sized particles (as in Fig. 1.8) and a much more compact film.

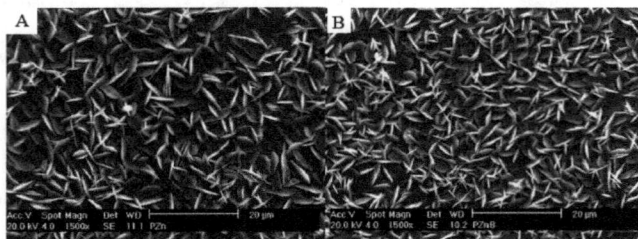

Fig. 1.8. *SEM images of phosphate coatings: (A) in PZn, (B) in PZn+BTAH.*

Another research by *F. Fang et al.* (2010) involving hydroxylamine sulfate (HAS) as accelerator. Added HAS reduces the process stages (amorphous precipitation, anodic depolarization and film thickening) by 53%, 31% and 50%, respectively; in time, the size of the crystals decreased from 100 to 50 μm, whereas the $Zn_2Fe(PO_4)_2 \cdot 4H_2O$ concentration increases from 30 to 44% (Fig. 1.9). HAS may be used in low temperature phosphating in order to replace traditional nitrate, being less polluting and having better phosphating rates, as well as superior film quality.

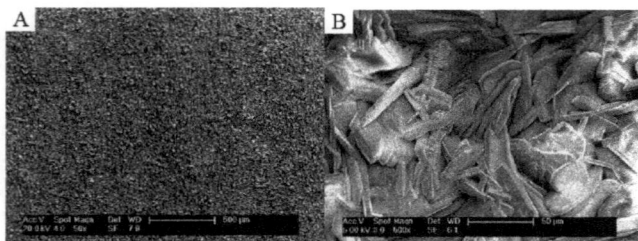

Fig. 1.9. *SEM images of the films: (A) X50, (B) X500.*

B. Narayanasamy et al. (2005) suggested a solution containing 150 ppm disodium phosphate (DSHP) and 50 ppm Zn^{2+}, with 97% efficiency as inhibitor controlling carbon steel corrosion. A synergic effect occurs between DSHP and Zn^{2+} and they work as a mixed inhibitor. The protective film consists of ferrous phosphate and $Zn(OH)_2$.

1.4.4   Zinc phosphates with specific color

G. Li et al. (2004) obtained a black phosphate film used on steels (Fig. 1.10). The solution includes ion phosphates (11.5 g/L – by the addition of 85% phosphoric acid), zinc ions (1.2 g/L – by zinc oxide addition), nitrate ions (0.5 g/L – by nitrate acid and manganese nitrate addition), manganese ions (4.0 g/L – by manganese nitrate addition) and also sodium molybdate as additive.

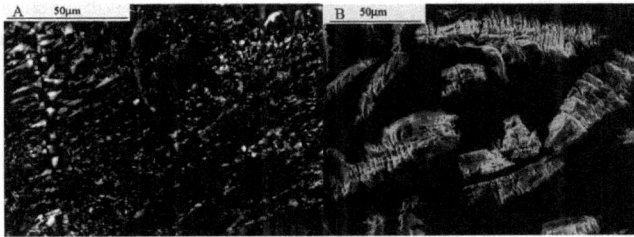

Fig. 1.10. *SEM images of the films: (A) 1000X black phosphate and (B) black phosphate without molybdate added.*

They concluded that even and finer grains are obtained as compared to traditional phosphating. The thickness is approximately 18 microns and the film has excellent corrosion resistance. Black phosphate may also be used as a lubricating layer *[Li, 2004]*.

A zinc phosphate solution containing at least 2% magnesium and 0.01…1% Ni, Co and Cu produces a very even white film *[Ishizuka, 2003]*.

1.4.5   The role of post-treatments

A post-treatment with sodium silicate solution was used to improve corrosion resistance of phosphate films *[Lin, 2008]*.

The corrosion protection effect of the phosphate film is improved by a sodium silicate treatment (Fig. 1.11). It is comparable to the one achieved by chromium post-treatments *[Lin, 2008; Seidel, 1995]*.

Materials Research Forum LLC
doi: http://dx.doi.org/10.21741/9781945291913

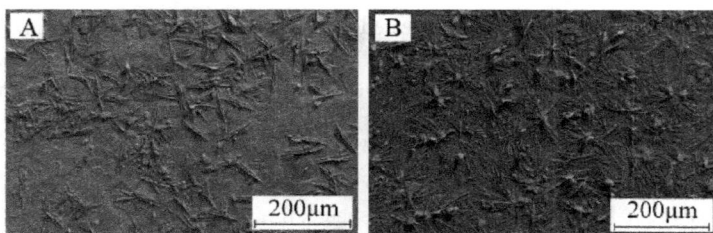

Fig. 1.11. SEM images of the phosphate layers obtained in
solutions after 30 s (A) and 300 s (B).

Another post-treatment consists of the use of a diluted poly-4-vinylphenol solution or reaction product between aldehyde or ketone and poly-4-vinylphenol [*Lindert, 1983*]. Surface conversion is improved for paint adhesion and corrosion protection properties.

Chapter 2

Creation of Thin Phosphate Films in the Laboratory

Given the expertise that we have acquired during the "Modern High Porosity Phosphating Coating Technology Used for the Volumic Plastic Working of Automotive Industry Parts" project contracted by the Partnership Program in priority areas, according to the contract no. 71-086/18.09.2007, developed between 2007 and 2009, and based on PhD thesis research conducted between 2009 and 2012, I chose a set of compatible materials involved in competitive surface coprecipitation processes that have not been previously studied and that are completely new and original as research topics.

It is a well-known fact that acid-base processes, accompanied by redox and complexing reactions, rely on competitive reactions which are difficult to control. Therefore, a small number of compatible cations were selected from various series of elements of the period table, which, by their expected outcome, would allow the representation of the whole series or group. A set of surface additives, which have not been studied so far, were also selected as high dispersion wetting agents.

The behavior of certain metal atoms or cations in different reactive systems with intermetallic compounds, zinc, titan, nickel and cobalt etc. oxides [*Bosinceanu, 2011; Iacomi, 2011, Pascariu, 2012, Pinzaru, 2011, Poiana, 2012a and 2012b; Tănase, 2010, 2011 and 2012, Vlad, 2012*] has allowed the selection of compatible systems in phosphating processes by acid aqueous media conversion susceptible to form even and compact film-forming layers adherent to the iron-based metallic substrate.

Therefore, this chapter describes the characteristics of the substrate used for precipitation deposits from acid aqueous solutions of metallic phosphates and of the processes involved in the creation of thin films with many practical uses, as the resistance to corrosion and lubrication were the focus of attention.

In this chapter we present new methods for thin phosphate layers obtained by conversion and the elaboration of the technological flow for microcrystalline phosphtation of iron objects by additive/substractive system. We used two systems, one with zinc cations and with metallic zinc.

A second method of crystalline lubricant phosphtation of metallic parts based on iron on substractive/additive system is presented with the elaboration of the technological flow for the crystalline lubricant phosphtation of iron objects on substractive/additive system.

The chemical composition and the parameters of the procedures are presented with the laboratory installation used for chemical crystalline phosphtation.

2.1   Experimental Protocol Development

Since the technologies employed to obtain thin surface films on metal surfaces by chemical precipitation involve several complex aspects related to material reactivity, to the chemical equilibriums established between competitive processes, and to the work parameters and characteristics of the substrate surfaces, they required the development of an experimental protocol that would involve the following aspects: material selection (support, reactants and additives), working condition selection, technological flow enhancement on stages together with the related processes/operations, selection of particular methods with related thin film examination techniques, by the particular involvement of coassistance or corroboration methods (microscopy with chemical, spectral, etc. analysis).

As concerns the thin film development procedure, we chose to immerse the metal specimens in acid aqueous solutions, as discs with different diameters (Fig. 2.1). By nucleation processes, these solutions are able to form poorly soluble phosphate and/or metallic nitrate films on metal surfaces. As for the working parameters, we should mention the following: *nature and concentration of the reactive components and additives, temperature, pH, stirring rate and immersion time.*

Adequate methodology has been used in the experiments conducted in the laboratory, in compliance with industrial technologies, with absolutely necessary stages and with the observance of safety standards.

Laboratory experiments include the following stages and related operations, which make up the ***technological flow***:

- manufacture of steel blanks;
- alkaline chemical degreasing;
- washing;
- acid chemical pickling;
- washing;
- phosphating;
- washing;
- use of phosphate-coated blanks in an oven at the set parameters;
- specimen sampling for phosphate film analysis;
- characterization of deposited films (OM, SEM, XRD).

2.2 Support Material Used for the Research (Blanks)

The support used for phosphate film deposition is a soft DC 01 steel which complies with the Carbon Steel AISI 1010 (SR EN 10130) American Standard, employed for plastic deformations, which has the composition described in table 2.1.

Table 2.1. *Composition of used steel.*

Element	Percentage	Element	Percentage
Fe	rest	Nb	0.001
C	0.045	Al	0.040
Si	0.024	Cu	0.039
Mn	0.239	As	0.002
P	0.012	Ca	0.002
S	0.013	Pb	0.005
Cr	0.003	Sn	0.001
Mo	0.003	Ti	0.001
Ni	0.29		

The DC 01 steel specimens were analyzed in the "Laboratory for expert's appraisal of materials by optical emission spectrometry" in the Center for Eco-metallurgical Research and Expert Appraisal of the Polytechnical University of Bucharest.

It is equipped with a GNR metalLab 75-80 optical emission spectrometer with a spectral measuring range between 120 and 800 nm. The spectrometer software allows the performance of quantitative analysis to determine the chemical composition of iron-based metallic alloys, as follows: C, Si, Mn, P, S, Cr, Mo, Ni, Nb, Al, Cu, Co, Ti, V, W.

Table 2.2 describes the mechanical characteristics of steel.

Table 2.2. *Mechanical characteristics of steel.*

Steel symbol	Standard	Mechanical characteristics				
		R_c, N/mm^2	R_m, N/mm^2	A_{80}, % min	r_{90}, min	n_{90}, min
DC 01	SR EN 10130+A1	-/280	270/410	28	-	-

The part roughness ranges from 0.3 $\mu m < R_a < 0.9$ μm.

A small-sized part was used to experiment modern phosphate coating technologies in laboratory conditions.

Materials Research Forum LLC
doi: http://dx.doi.org/10.21741/9781945291913

The specimens used are round, have 20 mm and 48 mm in diameter, respectively, and are 0.5 mm thick, as shown in figures 2.1 and 2.2.

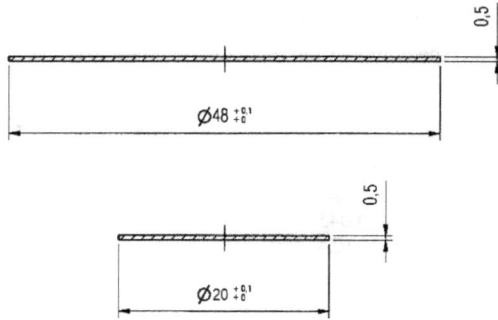

Fig. 2.1. *Size of steel specimens.*

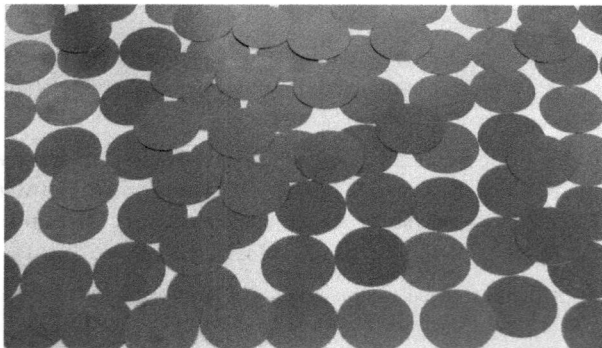

Fig. 2.2. *Test specimens.*

2.3 Experimental Equipment

The equipment used for phosphate coating included 4 thermostat baths with 200x400x200 mm tanks (Fig. 2.3).

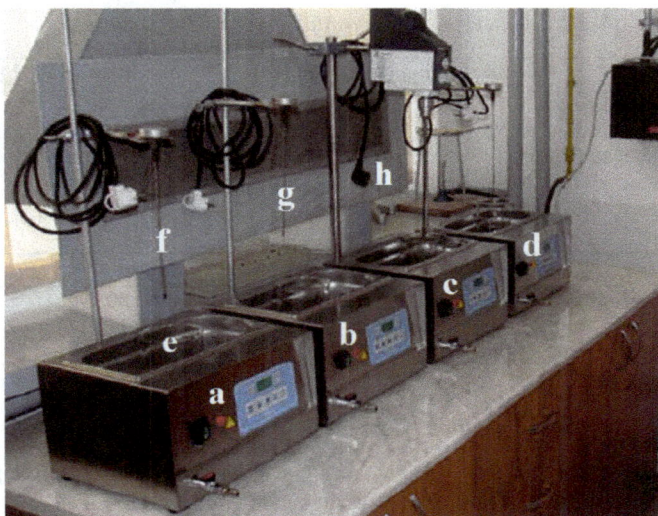

Fig. 2.3. *Phosphate-coating equipment: a, b, c and d – thermostat baths, e – laboratory beaker, f, g and h – stirrers.*

Each bath contains a stirrer (Fig. 2.3f-g-h), which prevents solution separation/sedimentation during the reactions.

A laboratory beaker is placed inside each bath (Fig. 2.3e) in order for it to be heated in the bain-Marie system. The first tank is used for degreasing (Fig. 2.3a), the second for pickling (Fig. 2.3b), and the third (Fig. 2.3c) and fourth (Fig. 2.3d) for the phosphating solutions, when a two-staged procedure is used.

2.4 Solution Preparation

2.4.1 *Degreasing solution*

Table 2.3. shows the chemical composition and parameters of the working parameters set for the alkaline chemical degreasing of steel parts by immersion in a stirring tank, according to figure 2.4.

Table 2.3. *Chemical composition and working parameters*

Chemical components	Concentration, [g/L]
Sodium hydroxide, NaOH	40
Sodium carbonate, Na_2CO_3	30
Trisodium phosphate, $Na_3PO_4 \cdot 10H_2O$	30
Sodium silicate, $Na_2SiO_3 \cdot 9H_2O$	5
Detergent (surfactant)	3...10
Working parameters	**Value**
Temperature, [°C]	80...90
pH	11...12
Degreasing time, [min]	10

2.4.2   Chemical pickling solution

The chemical composition and working parameters used for acid chemical pickling of steel parts in static tank by immersion procedure is shown in Table 2.4.

Table 2.4. *Chemical composition and working parameters.*

Chemical components	Concentration, [g/L]
Hydrochloric acid, HCl (ρ=1.19g/cm^3)	150
Hexamethylenetetramine, $C_6H_{12}N_4$	0.45
Sodium sulfate decahydrate, $Na_2SO_4 \cdot 10H_2O$	0.15
Working parameters	**Value**
Temperature, [°C]	20...25
pH	1...2
Pickling time, [min]	max. 30

Figures 2.4 and 2.5 show the involvement of tanks used for degreasing and pickling, respectively, in the experimental laboratory equipment.

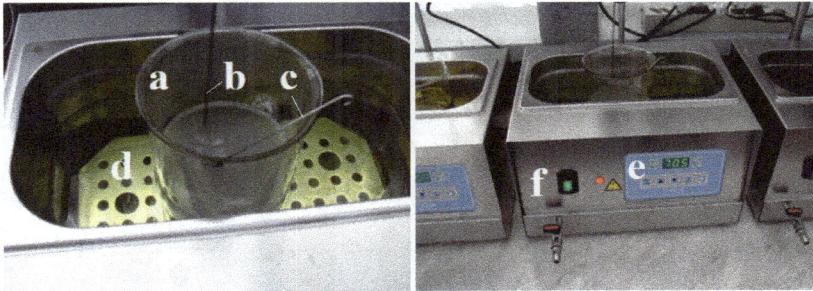

Fig. 2.4. *Tanks used for alkaline chemical degreasing: a – laboratory beaker, b – stirrer, c – specimen holder, d – heating element, e – thermostat, f – on/off switch.*

Fig. 2.5. *Tanks used for acid chemical pickling: a – laboratory beaker, b – stirrer, c – specimen holder, d – thermostat, e – on/off switch.*

2.4.3 Phosphating solution used as a reference

Table 2.5. shows the components and working parameters of the standard phosphating solution.

Table 2.5. Components and working parameters of the basic solution.

Chemical components	Quantity
H_3PO_4 85% ($\rho = 1.68g/cm^3$)	8.16 mL
Zn splinters	4 g
HNO_3 65% ($\rho = 1.40g/cm^3$)	2.6 mL
NaOH	0.75 g
$NaNO_2$	0.45g
$Na_3P_3O_{10}$	0.05g

Working parameters	Value
Temperature, [°C]	80...90
pH	< 3.50
Time, [min]	30

2.4.4 Experimental solutions

Either the chosen reactive components, as such or combined, or the chosen surfactants, as the case may be, were added to the stock solution.

Thus, table 2.6. shows the components that were added to the stock solution.

Table 2.6. Components involved in experiments.

Specimen	Chemical component	Quantity per Liter
2A + 2B	Hexamethylenetetramine (0.2% alcoholic solution)	2g
3A + 3B	Thiourea (99%)	10g
4A + 4B	Oak tannin (100%)	10g
X1	$CoCl_2 \cdot 6H_2O$	12g
X2	$Ni(NO_3)_2 \cdot 6H_2O$	12g
X3	CrO_3	100g
X4	$Bi(NO_3)_3 \cdot H_2O$	20g
Y1	$Ni(HCOO)_2 \cdot 2H_2O$	12g
Y2	$MgS_2O_3 \cdot 6H_2O$	16g
Y3	$Mn(NO_3)_2 \cdot 4H_2O$ (partly soluble in water of crystallization)	120g
Y4	$Mg(NO_3)_2 \cdot 6H_2O$	15g

During the process of creation of thin phosphate films by using hexamethylenetetramine, thiourea and tannin additives, we noticed that the thin films deposited further to the immersion of the specimens in the standard solutions with additives were uneven and thin, and the crystallites were heterogeneous. These processes are due to the formation of complex combinations of additives and precipitation cations, which do not allow phosphate development on steel surfaces.

In these circumstances, the phosphating process was conducted differently for the first three sets of specimens and it consisted of initial phosphating (15 min) in the stock solution, followed by secondary phosphating in solution with added hexamethylenetetramine, thiourea and oak tannin (15 min). This approach was resorted to since it was noticed that solutions with additives does not support primary nucleation on steel surfaces. Their role becomes conspicuous after the occurrence of the first poorly soluble crystalline grains, which support certain aspects of nanostructure morphology and distribution.

An ADAM AFP 400 analytical balance with four decimals was used for material dosing for the solutions described above.

Since the baths had thermostats, the temperature was monitored automatically.

Figure 2.6 shows the specimen holders used during the degreasing, pickling and phosphating processes.

After they have been rinsed with distilled or deionized water, the specimens were immersed in phosphating baths for 30 minutes; the 80...90°C baths were continuously stirred at 500 rpm.

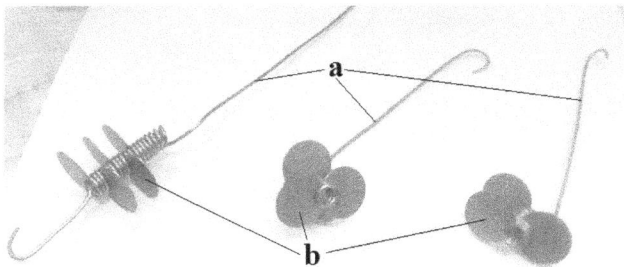

Fig. 2.6. *Specimen holders (a) and test specimens (b).*

Stock solution preparation included the following stages:

- zinc splinter dissolution in the orthophosphoric acid solution; a white-grayish dense paste is obtained (fig. 2.7a);
- adding nitric acid solution to the dispersion described above (fig. 2.7b);

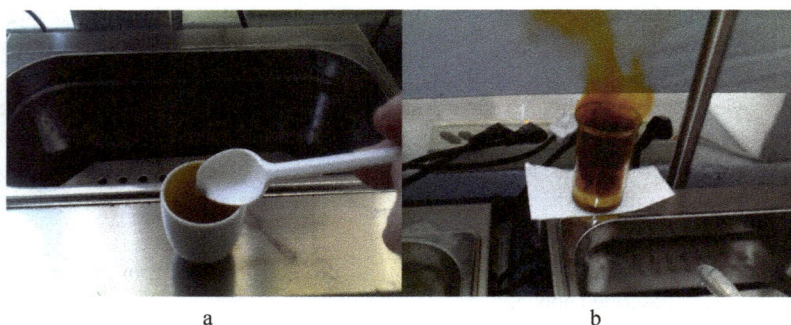

a   b

Fig. 2.7. *Dissolution under the hood of zinc splinters in the orthophosphoric acid solution (a) and addition of nitric acid solution with nitrogen dioxide (NO_2) release (b).*

- separately, the sodium hydroxide, sodium nitrite, and sodium și trisodium phosphate are dissolved in 100mL of distilled water;
- the system of the two acid solutions is dispersed in this solution, after the reaction has finished (the nitrogen dioxide release ceases);
- distilled water is poured up to 1L and gently stirred.

The components involved in the experiment are added to this stock solution (table 2.6), when colorful solutions are obtained depending on the metallic cation or additive used (Fig. 2.8.).

Fig. 2.8. *Color of solutions with additives for experiments:*
a – $CoCl_2$, b – $Ni(NO_3)_2$, c – CrO_2, d - $Bi(NO_3)_3$,
e - $Ni(HCOO)_2$, f - MgS_2O_3, g - $Mn(NO_3)_2$, h - $Mg(NO_3)_2$.

Figure 2.9 shows the phosphating tanks during the thin film deposition processes.

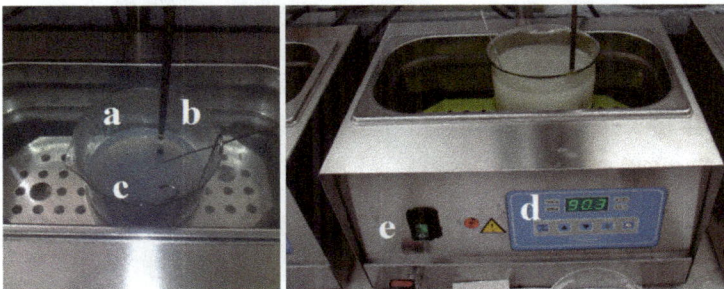

Fig. 2.9. *Tanks during phosphating: - laboratory beaker, b – stirrer,*
c – specimen holders, d – thermostat, e – on/off switch.

After phosphating, the specimens were rinsed and dried in an oven at 105±5°C, for 4 hours (Fig. 2.10), then they underwent laboratory analyses and tests: corrosion test, tribology tests and plastic cold working.

Fig. 2.10. Parts phosphate-coated in a drying oven.

Other specimens were later sampled for certain tests, in order to obtain cross sections, by embedding in epoxy resin and grinding.

Figure 2.11 shows embedded and ground specimens, prepared for microscopic analysis in cross sections.

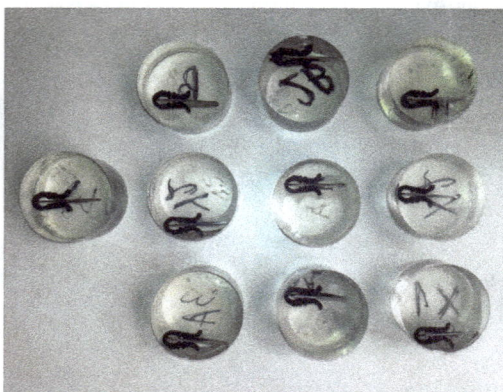

Fig. 2.11. Test specimens embedded in epoxy resin for analysis in cross section.

2.5 Modeling of the Structure of the Phosphate Coating from the Viewpoint of the Surface Processes

As shown in chapter I, the chemical phosphating processes include an activation phase, in which the surface of the metal acts as a poly-electrode, with the formation of short-circuited micropiles, made up of the two areas: anodic, where the metal is dissolved, and cathodic, where hydrogen is released. These two reactions may occur simultaneously or successively, leading to the formation of discrete polarization areas with electron transfer [*Rausch, 1990; Crow, 1994; Grünwald, 1995; Marcus, 2002*].

The anodic oxidation process (Fe $\rightarrow$ Fe^{2+} + 2e$^-$) may be favorably displaced towards the right in the presence of oxidants (NO$_3^-$, Cr$_2$O$_7^{2-}$, MoO$_4^{2-}$, WO$_4^{2-}$ etc.), which act as electron and proton consumers.

The development of the process in the sense of the formation of a poorly soluble tertiary phosphate film depends on the existence of well-defined relations between the precipitation compounds formed, the various ions and other molecular species in the solution, which, by competitive reactions, determine free phosphoric acid consumption or formation.

Here is the reaction specific to tertiary phosphate film formation from primary phosphate film:

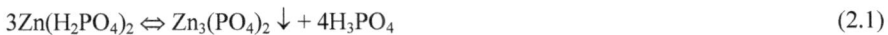

$$3Zn(H_2PO_4)_2 \Leftrightarrow Zn_3(PO_4)_2 \downarrow + 4H_3PO_4 \tag{2.1}$$

Similarly, in the anodic area, the Fe^{2+} ion initially reacts with the dihydro-phosphate anion, thus forming primary salt:

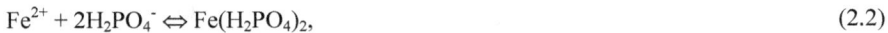

$$Fe^{2+} + 2H_2PO_4^- \Leftrightarrow Fe(H_2PO_4)_2, \tag{2.2}$$

and the Fe^{2+} ion excess displaces the equilibrium of the primary into tertiary salt conversion reaction, leading to the formation of Zn(II) and Fe(II) ortho and pyrophosphate, where the double phosphate tetrahydrate dominates - *phosphophyllite* (Zn$_2$Fe(PO$_4$)$_2$·4H$_2$O) [*Sandu, 2012*], formed according to the reaction:

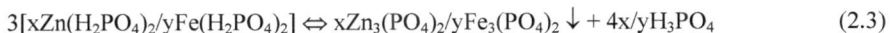

$$3[xZn(H_2PO_4)_2/yFe(H_2PO_4)_2] \Leftrightarrow xZn_3(PO_4)_2/yFe_3(PO_4)_2 \downarrow + 4x/yH_3PO_4 \tag{2.3}$$

The modeling of the phosphating processes according to the reactions described above includes three stages, summarized in figure 2.13.

According to the chart, after the initial primary phosphate formation, nitric acid is added to the resulting white-grayish dispersion, the role of which is to displace the equilibrium of the anodic oxidation reaction and of the tertiary phosphate formation reaction at the surface of the metal, by activating the competitive phosphophyllite obtaining processes, under the form of a dendritic intergrown congruent in addition to zinc phosphates and other metallic cations. These processes are modeled by the presence of surfactants.

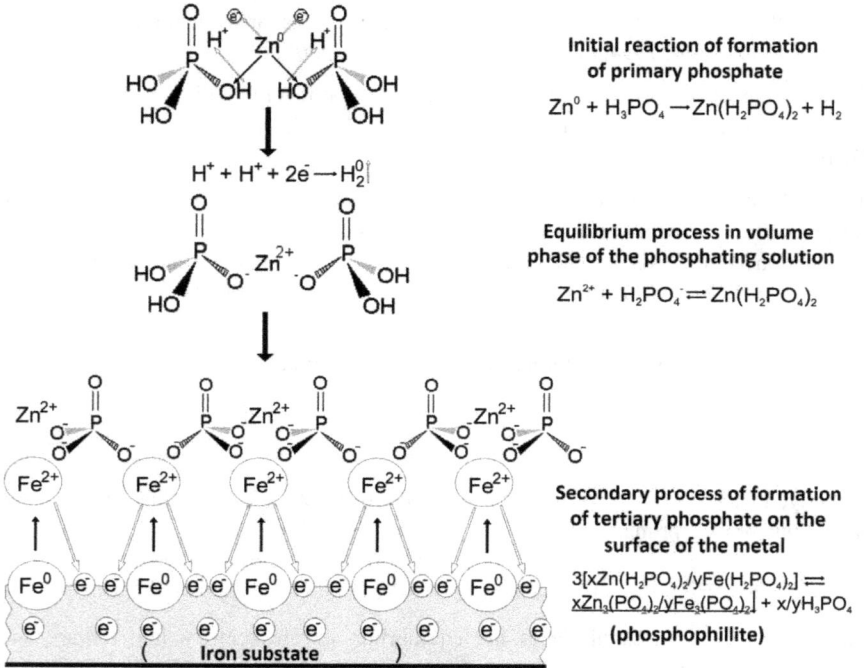

Fig. 2.13. Chart of the film-forming phosphating process.

Chapter 3

Modern Procedures for the Creation of Thin Anticorrosive Lubricating Phosphate Films by Addition and Subtraction

The microcrystalline phosphating procedures applying to iron-based metallic parts, which have already been patented, focused on the involvement of competitive surface processes that are able, through differential nucleation, to confer thin films high corrosion protection and load bearing capacity, in order to allow the insertion/embedding of solid lubricating microstructures with multiple improvement actions of the processing characteristics and implicitly of the protection provided in the severe plastic deformation processes.

The great advantage was the involvement of the authors in an interdisciplinary research team, on a particular topic in the field, which was extremely attractive to us and enabled us to improve our theoretical and technological knowledge.

The important data gathered on this occasion allowed the identification of innovative materials and processes used to created phosphate thin films with multiple uses.

It is well known that for the creation of thin films by sequential coprecipitation of poorly soluble salts of the orthophosphate, pyrophosphate and/or nitrate ions, with various metallic cations, which have an anticorrosive, lubricating and also esthetic role, one uses iron surface passivation procedures, which rely on acid-base processes, accompanied by redox and complexing processes.

According to literature data [*Oniciu, 1980 and 1982; Marinescu, 1984; Rausch, 1990; Crow, 1994; Grünwald, 1995; Schlesinger, 2000; Sandu, 2011 and 2012a*], after the prior cleaning of the parts by classical degreasing and pickling procedures, these procedures rely of part surface treatment processes using solutions which contain orthophosphoric acid, accompanied by nitric acid; even compact and substrate-adherent films, poorly soluble in acid environments are thus formed; they consist of coatings of orthophosphates, pyrophosphates and nitrates of certain metals, in stable oxidation states, in the presence of additives or surfactants like: polyacrylamide, epoxy esters, silicates, citric acid, sulfamic acid, etc.

The disadvantage of these procedures is the fact that the films created are very thin, almost transparent, and uneven.

3.1 New Phosphating Procedures by Conversion as Thin Surface Films

The microcrystalline phosphating procedure of iron-based metallic parts, which consists of chemical passivation of steel and cast iron part surfaces [*Bejinariu, 2011a and b*], resolves the disadvantages listed above. In other words, in order to obtain thin and highly porous surface films, capable of allowing the insertion/embedding of lubricating solid structures, with multiple improvement actions of the processing characteristics and implicitly of the protection provided, and with good adherence to iron-based metallic substrates, a two-stage chemical sequential treatment is applied after degreasing and pickling:

- a thin porous copper film obtained by coating (3.1) or a nickel film obtained by electrodeposition (3.2),

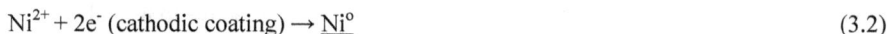

$$\underline{Fe^\circ} + CuSO_4 \rightarrow \underline{Cu^\circ} + FeSO_4, \tag{3.1}$$

$$Ni^{2+} + 2e^- \text{ (cathodic coating)} \rightarrow \underline{Ni^\circ} \tag{3.2}$$

- followed by nickel or copper pyrophosphate precipitation, by part immersion in two acid aqueous solutions, one which is slightly oxidizing and based on phosphate anion and Zn^{2+} cations, in the presence of permanganate anion, as surface moderator (3.3), and the other based on orthophosphoric acid, which contains fine metallic zinc powder (3.4), which is differentially applied by immersion, at room temperature, on the cleaned surfaces,

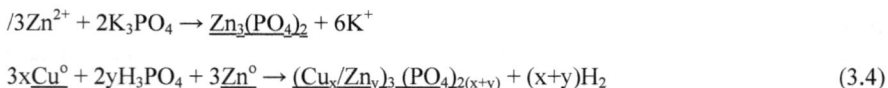

$$15\underline{Cu^\circ} + 6KMnO_4 + 16H_3PO_4 \rightarrow$$

$$\underline{5Cu_3(PO_4)_2} + 2\underline{Mn_3(PO_4)_2} + 2K_3PO_4 + 24H_2O/ \tag{3.3}$$

$$/3Zn^{2+} + 2K_3PO_4 \rightarrow \underline{Zn_3(PO_4)_2} + 6K^+$$

$$3x\underline{Cu^\circ} + 2yH_3PO_4 + 3\underline{Zn^\circ} \rightarrow \underline{(Cu_x/Zn_y)_3\,(PO_4)_{2(x+y)}} + (x+y)H_2 \tag{3.4}$$

Thus, the procedure relies on the phosphating in an acid environment process by an **additive/subtractive** mechanism, in the presence of the Cu, Zn and Mn cations, which are stable in the oxidation state (II) and which, after precipitation, become inert in relation to the Fe(0) in the substrate.

These cations are only susceptible to acid-base processes, when thin porous films are formed by coprecipitation as orthophosphates, even in the presence of poorly hydrated Fe(II, III)-based oxidic or saline stains, which are solubilized by light surface oxidation in the presence of permanganate anion.

The film thus formed is then dipped in graphite or molybdenite colloidal hydroalcoholic dispersions, which have undergone steric and electrostatic stabilization, in the presence of the NH_4OH-NH_4Cl buffer system, which ensures an optimum pH of 8.5...9.00.

This procedure has a series of advantages:

- it is easily applied, by dipping, at normal temperatures, without any energy consumption;
- it allows the creation by synergy of a thin porous structure, which does not interfere with part dipping in lubricant colloidal systems;
- it may be applied on any iron-based metallic substrate, such as cast iron and steel, as finite cast parts, which have undergone plastic volumic or cold surface working;
- the films formed are highly reliable and stable in time.

Here are three variants of the procedure.

Variant 1 of the procedure

Regardless of the type of the part, one should begin by preparing the metallic surfaces, by means of the classical degreasing and pickling methods. When the surfaces are covered with coarse coatings, scale and burr, these will be removed by sandblast cleaning, brush cleaning or other mechanical procedures.

Aqueous solution systems are used for degreasing, which include: sodium hydroxide, sodium carbonate, trisodium phosphate and sodium silicate, for 10 minutes at 80...90°C temperature, with a pH ranging between 11 and 12.

An aqueous solution made up of the following elements is used for oxidic and saline film pickling: hydrochloric acid, hexamethylenetetramine and sodium sulfate decahydrate, for 30 minutes at room temperature.

The degreased and pickled parts undergo a two-stage sequential chemical treatment:

- first a thin porous copper film is deposited by cementation or a thin porous nickel film is deposited by electrodeposition; thus, the parts are immersed for 3...5 min in a slightly acid aqueous $CuSO_4$ solution, containing 1...3% tannin (synthetic or natural), obtained by the dissolution of 125 grams of copper sulfate pentahydrate in 1 liter of distilled water (0.5M of Cu^{2+} solution) in which 10...30 g of natural or synthetic tannin were previously dissolved, then the solution thus obtained was slightly acidulated by using a few drops of H_2SO_4 10%, the result being a fine spongy film of amorphous metallic copper;

- the part covered by this film is then immersed for 30 minutes in aqueous solution of 2M orthophosphoric acid, in the presence of 0.1M permanganate anion and of Zn^{2+} 1M cation, slightly stirred. After this precipitation process, the part undergoes several successive rinsing operations in distilled water, by 10 minute immersion, until the pH of the rinsing solution becomes constant, then the parts are dried in a thermally regulated oven, at 110±5°C temperature, for 20 minutes.

After drying, the parts are immersed in 10% graphite or colloidal molybdenite hydroalcoholic dispersions, which have undergone a steric and electrostatic stabilization process, in the presence of a surfactant (for instance: sodium aryl sulfonate or methanesulfonic acid sodium salt) and of the NH_4OH-NH_4Cl buffer system, which contains 1.2% NH_4OH and 0.5% NH_4Cl, which confers the best insertion pH ranging from 8.5 to 9.

The film thus formed allows good workability by volumic plastic deformation or cold surface working, due to its lubricant power, and provides high protection, both during these operations and afterwards.

Variant 2 of the procedure

It resembles the previous application, no matter what the type of the part is; for starters, the metallic surfaces are prepared by degreasing and pickling.

The degreased and pickled parts are first subjected to a chemical sequential treatment, which also comprises two stages:

- a thin porous nickel layer is deposited by cathodic means;
- the part is then treated with an aqueous solution of 2M orthophosphoric acid, for 30 minutes, in the presence of 0.1M permanganate anion and of Zn^{2+} 1M cation, slightly stirred. After this precipitation process, the part undergoes several successive rinsing operations in distilled water, by 10 minute immersion, until the pH of the rinsing solution becomes constant, then the parts are dried in a thermally regulated oven, at 110 ±5 °C temperature, for 20 minutes.

After drying, the parts are immersed in 10% graphite or colloidal molybdenite hydroalcoholic dispersions, which have undergone a steric and electrostatic stabilization process, in the presence of a surfactant (for instance: sodium aryl sulfonate or methanesulfonic acid sodium salt) and of the NH_4OH-NH_4Cl buffer system, which contains 1.2% NH_4OH and 0.5% NH_4Cl, which confers the best insertion pH ranging from 8.5 to 9.

The film thus formed allows good workability by volumic plastic deformation or cold surface working, due to its lubricant power, and provides high protection, both during these operations and afterwards.

Variant 3 of the procedure

Just like in the previous cases, the iron-based metallic parts are prepared by degreasing and pickling.

The parts are then immersed for 3 to 5 minutes in a slightly acid aqueous $CuSO_4$ solution, containing 1...3% synthetic or natural tannin, obtained by the dissolution of 125 grams of copper sulfate pentahydrate in 1 liter of distilled water (0.5M of Cu^{2+} solution) and 10...30 g of tannin, which is then slightly acidulated by using a few drops of H_2SO_4 10%, the result being a fine spongy film of amorphous metallic copper.

The part thus coated is immersed for 30 minutes in aqueous solution of 2M orthophosphoric acid, which contains 50 g of fine metallic zinc powder dispersed in 1 liter of solution.

After this precipitation process, the part undergoes several successive rinsing operations in distilled water, by 10 minute immersion, until the pH of the rinsing solution becomes constant, then the parts are dried in a thermally regulated oven, at 110 ±5 °C temperature, for 20 minutes.

After drying, the parts are immersed in 10% graphite or colloidal molybdenite hydroalcoholic dispersions, which have undergone a steric and electrostatic stabilization process, in the presence of a surfactant (for instance: sodium aryl sulfonate or methanesulfonic acid sodium salt) and of the NH_4OH-NH_4Cl buffer system, which contains 1.2% NH_4OH and 0.5% NH_4Cl, which confers the best insertion pH ranging from 8.5 to 9.

The film thus formed allows good workability by volumic plastic deformation or cold surface working, due to its lubricant power, and provides high protection, both during these operations and afterwards.

To conclude with, the following statements stand true for the crystalline chemical phosphating procedure in the **additive/subtractive** system:

a) the microcrystalline phosphating procedure applied to iron-based metallic parts is aimed at obtaining thin and highly porous surface films, which allow the insertion/embedding of solid lubricant structures with multiple improvement actions of the processing characteristics and implicitly of the protection provided, with good adherence to the iron-based metallic substrates; after the classical degreasing and pickling operations, the iron-based parts undergo a two-stage sequential chemical treatment: first a thin porous copper film is deposited by coating or a thin porous nickel film is deposited by electrodeposition, followed by nickel or copper pyrophosphate precipitation, by part immersion in two acid aqueous solutions, one which is slightly oxidizing and based on phosphate anion and Zn^{2+} cations, in the presence of permanganate anion, as surface moderator and the other based on orthophosphoric acid, which contains fine metallic zinc powder differentially applied by immersion, at room temperature;

b) the specificity of the procedure consists of the fact that, in order to obtain thin porous films, an acid-base process is applied together with a redox procedure by the additive/subtractive mechanism; also, a film may also be created by Cu^0 deposition, after part immersion in a slightly acid aqueous $CuSO_4$ solution obtained by the dissolution of 125 grams of copper sulfate pentahydrate in 1 liter of distilled water (0.5M of Cu^{2+} solution) and 10...30 g of natural or synthetic tannin slightly acidulated by using a few drops of H_2SO_4 10%; the part is then immersed for 30 minutes in an aqueous solution of 2M orthophosphoric acid, in the presence of 0.1M permanganate anion and of Zn^{2+} 1M cation, slightly stirred. Then, the part undergoes several successive rinsing operations by 10 minute immersion in distilled water, until the pH of the rinsing solution becomes constant, followed by drying in a thermally regulated oven, at 110 ±5 °C temperature, for 20 minutes;

c) in order to obtain thin porous phosphate films, a thin porous nickel film is deposited by cathodic means; the part is then treated by 30 minute immersion in an aqueous solution of 2M orthophosphoric acid, in the presence of 0.1M permanganate anion and of Zn^{2+} 1M cation, slightly stirred; then, the part undergoes several successive rinsing operations by 10 minute immersion in distilled water, until the pH of the rinsing solution becomes constant, and it is finally dried in a thermally regulated oven, at 110 ±5 °C temperature, for 20 minutes;

d) after the thin porous copper phosphate- or nickel-based films have been deposited, the part is immersed for 30 minutes in an aqueous solution of 2M orthophosphoric acid, which contains 50 g of fine metallic zinc powder dispersed in 1 liter of solution; this is followed by successive rinsing operations by 10 minute immersion in distilled water, until the pH of the rinsing solution becomes constant; the part is then dried in a thermally regulated oven, at 110 ±5 °C temperature, for 20 minutes;

e) in order to allow the insertion of a micro-heterogeneous graphite or molybdenite system, which is designed to increase its lubricant power, the part is immersed in a 10% graphite or colloidal molybdenite hydroalcoholic dispersion, which has undergone a steric and electrostatic stabilization process, in the presence of a surfactant and of the NH_4OH-NH_4Cl buffer system, which contains 1.2% NH_4OH and 0.5% NH_4Cl, which confers the best insertion pH ranging from 8.5 to 9.

3.2 Development of the Technological Flow Applying to Microcrystalline Phosphating of Iron-Based Metallic Parts Involving the Additive/Subtracting System

3.2.1 Additive/Subtractive Systems Based on Zinc Cations

The technological procedure of microcrystalline phosphating of ferrous metallic parts in the **additive/subtractive** system based on zinc cations includes several stages and phases shown in figure 3.1, which occur in the following sequence:

➢ obtaining steel blanks;
➢ alkaline chemical degreasing;
➢ cold running water rinsing;
➢ acid chemical pickling;
➢ cold running water rinsing;
➢ additivation by deposition according to reaction (3.1);
➢ control specimen sampling for film testing to optimize variable parameters;
➢ zinc cation-based subtractive phosphating according to reaction (3.3);
➢ cold water rinsing;
➢ phosphate-coated blanks dried in the oven at the set parameters;
➢ phosphate-coated control specimen sampling for phosphate film testing;
➢ graphite/molybdenite-coating;
➢ graphite/molybdenite-coated blank drying;
➢ graphite/molybdenite-coated control specimens sampled for testing to optimize the working parameters;

> plastic working of phosphate- and graphite/molybdenate-coated blanks;
> phosphate-coated control specimen sampling for phosphate film testing after plastic deformation;
> finite parts.

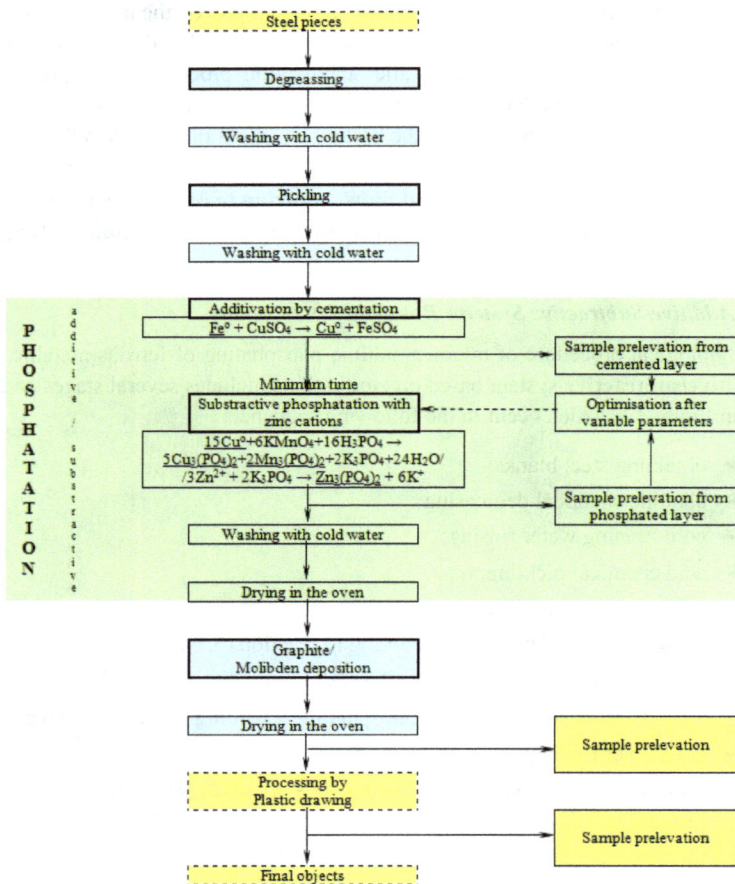

Fig. 3.1. *Technological flow of microcrystalline phosphating of ferrous metallic parts in the additive/subtractive system based on zinc cations.*

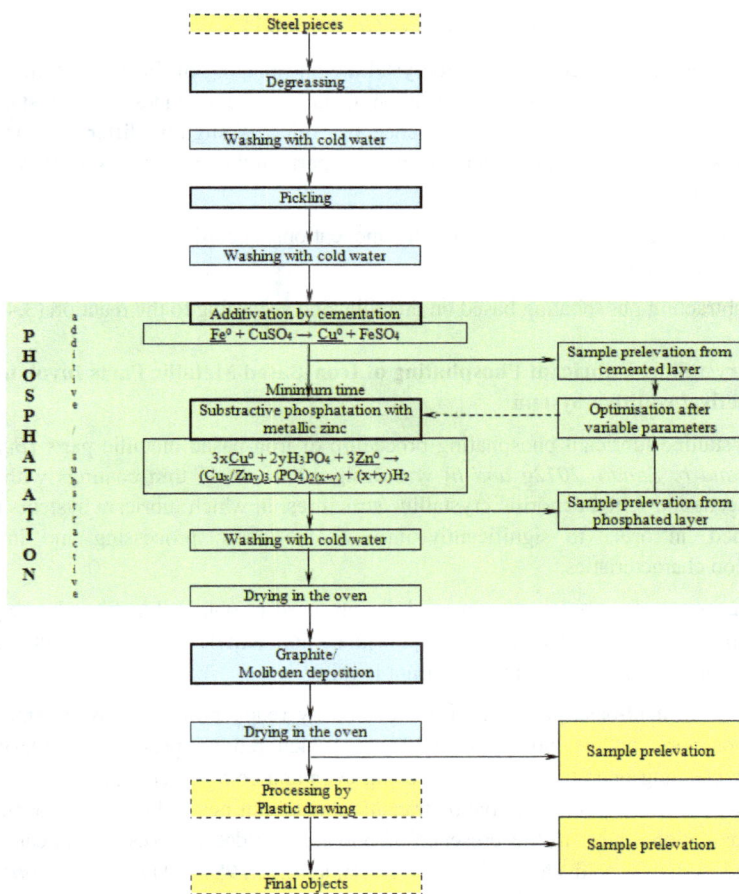

Fig. 3.2. *Technological flow of microcrystalline phosphating of ferrous metallic parts in the additive/subtractive system based on metallic zinc.*

3.2.2 Additive/Subtractive Systems Based on Metallic Zinc

The technological procedure of microcrystalline phosphating of ferrous metallic parts in the **additive/subtractive** system based on metallic zinc includes several stages and phases shown in figure 3.2., the sequence of which is slightly different than in the microcrystalline phosphating of ferrous metallic parts in the **additive/subtractive** system based on zinc cations:

- subtracting phosphating based on zinc cations according to the reaction (3.3), replaced by

- subtracting phosphating based on metallic zinc according to the reaction (3.4).

3.3 Crystalline Lubricant Phosphating of Iron-Based Metallic Parts Involving the Subtractive/Additive System

The crystalline lubricant phosphating procedure of iron-based metallic parts [*Bejinariu, 2011a and b; Sandu, 2012g and h*] was designed to obtain thin coatings with highly arborescent and felted dendritic crystalline structures, in which lubricant systems may be embedded, in order to significantly improve the parts' processing and implicitly protection characteristics.

Metallic iron surface passivation procedures are used to obtain thin films by sequential coprecipitaton of poorly soluble pyrophosphate ion and/or nitrate salts, and also to achieve corrosion protection and a pleasant look.

They rely on acid-base processes accompanied by redox processes, by treatment with orthophosphoric and/or nitric acid solutions, which require previous degreasing and pickling cleaning operations, the result being even and compact coatings adherent to the substrate, which are poorly soluble by transition metal nitrates and/or orthophosphates, in stable oxidation states, in the presence of polyacrylamide, epoxyesters, silicates, citric acid and sulfamic acid etc. [*Ghali, 1972; Askienazy, 1980; Oniciu, 1980 and 1982; Marinescu, 1984; Rausch, 1990; Grunwald, 1995; Gosset, 1989; Schlesinger, 2000; Ishii, 2001; Marcus, 2002; Sandu, 2011*].

The disadvantage of these procedures is the fact that the passivator film created is very thin, almost transparent, compact and often uneven, infested by oxidic stains formed *in situ* or induced after part commissioning, in the presence of humidity and the lubrication of which requires the application of superposed coatings, which are hardly adherent or have low carrying capacity.

The lubricant crystalline phosphating procedure applied to iron-based metallic parts by chemical passivation of steel and cast iron part surfaces neutralizes the disadvantages mentioned above, in the sense that it is designed to allow the creation of thin and highly porous surface coatings capable of permitting the insertion/embedding of solid lubricating structures with multiple improvement actions of the processing characteristics and implicitly of the protection, and with good adherence to iron-based metallic substrates.

After degreasing and pickling, a two-stage sequential chemical treatment is applied:

- iron pyrophosphate is first precipitated in the presence of the nitrate ion and of hydroxylamine sulfate (SHA) in acid environment, according to the reaction,

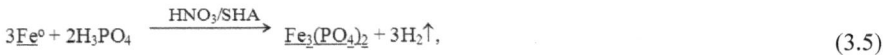

$$3\underline{Fe}^o + 2H_3PO_4 \xrightarrow{\text{HNO}_3/\text{SHA}} \underline{Fe_3(PO_4)_2} + 3H_2\uparrow, \tag{3.5}$$

- then the zinc pyrophosphate is inserted by coprecipitation, more precisely by immersing the parts in orthophosphoric acid which contains fine metallic zinc powder, at 90°C temperature, for 30 minutes, according to the reaction,

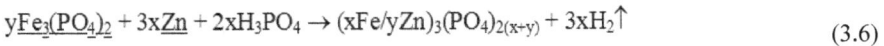

$$y\underline{Fe_3(PO_4)_2} + 3x\underline{Zn} + 2xH_3PO_4 \rightarrow (xFe/yZn)_3(PO_4)_{2(x+y)} + 3xH_2\uparrow \tag{3.6}$$

Thus, the procedure relies on phosphating in acid environment by a **subtractive/additive** mechanism with insertion, in the presence of the Zn^{2+} cations, which trigger even crystal increase at 90°C, thus forming highly arborescent and felted dendritic textures, with good retention power of colloidal lubricant suspensions in aqueous or organic systems.

The coating thus formed is them dipped in colloidal graphite or molybdenite hydroalcoholic dispersions, which have undergone steric and electrostatic stabilization, in the presence of the NH_4OH-NH_4Cl buffer system, which ensures an optimum pH of 8.5...9.00.

This procedure has certain disadvantages:

- it is easily applied, by immersion, at relatively low temperatures;
- it allows the achievement by synergy of a thin structure of highly arborescent and felted microcrystalline dendrites, which allow the embedment of lubricant colloidal systems with high carrying capacity;

- it may be applied on any iron-based metallic substrate, such as cast iron and steel, as finite or cast parts, parts that underwent volumic plastic deformation or cold surface working;

- the coatings thus formed have great reliability and stability in time.

We will describe hereunder a practical application of the procedure.

Variant of the procedure

Regardless of the type of the part, one should begin by preparing the metallic surfaces, by means of the classical degreasing and pickling methods. When the surfaces are covered with coarse coatings, scale and burr, these will be removed by sandblast cleaning, brush cleaning or other mechanical procedures.

Aqueous solution systems are used for degreasing.

Aqueous solutions are used for oxidic and saline film pickling.

The degreased and pickled parts undergo a two-stage sequential chemical treatment: iron pyrophosphate is first precipitated in the presence of the nitrate ion and of hydroxylamine sulfate (SHA) in an acid environment; then the zinc pyrophosphate is inserted by coprecipitation, more precisely by immersing the parts in orthophosphoric acid which contains fine metallic zinc powder, at 90 °C temperature, for 30 minutes.

Thus, a thin iron phosphate coating is formed by pyrophosphate ion reaction with metallic iron, in the presence of the nitrate ion and of hydroxylamine sulfate, at room temperature, after which the part is inserted in 20% orthophosphoric acid solution, which contains 50 g of fine metallic zinc powder dispersed in 1 liter of solution, at 90 °C temperature, for 30 minutes, when an interstitial zinc phosphate precipitation process occurs in the iron phosphate structure, by a subtractive/additive process.

After these precipitation processes the part undergoes several successive rinsing operations in distilled water, by 10 minute immersion, until the pH of the rinsing solution becomes constant, then the parts are dried in a thermally regulated oven, at 110 ± 5°C temperature, for 20 minutes.

After drying, the parts are immersed in 10% graphite or colloidal molybdenite hydroalcoholic dispersions, which have undergone a steric and electrostatic stabilization process, in the presence of a surfactant (for instance: sodium aryl sulfonate or methanesulfonic acid sodium salt) and of the NH_4OH-NH_4Cl buffer system, which contains 1.2% NH_4OH and 0.5% NH_4Cl, which confers the best insertion pH ranging from 8.5 to 9.

The film thus formed allows good workability by volumic plastic deformation or cold surface working, due to its lubricant power, and provides high protection, both during these operations and afterwards.

To conclude with, the following statements stand true for the crystalline chemical phosphating procedure in the **additive/subtractive** system:

- the lubricant crystalline phosphating procedure applied to iron-based metallic parts is aimed at obtaining thin surface coatings, with highly arborescent and felted dendritic crystalline microstructures, with high carrying capacity for microcolloidal lubricant systems, with good adherence to iron-based metallic substrates; after the classical degreasing and pickling operations, the iron-based parts undergo a two-stage sequential chemical treatment: iron pyrophosphate is first precipitated in the presence of the nitrate ion and of hydroxylamine sulfate (SHA) in an acid environment; then the zinc pyrophosphate is inserted by coprecipitation, more precisely by immersing the parts in orthophosphoric acid which contains fine metallic zinc powder, at 90 °C temperature, for 30 minutes; the interstitial zinc phosphate precipitation process occurs in the iron phosphate structure, by a subtractive/additive process; after the rinsing and drying processes, the parts are immersed in 10% graphite or colloidal molybdenite hydroalcoholic dispersions, which have undergone a steric and electrostatic stabilization process.

3.4 Development of the Technological Flow of Crystalline Lubricant Phosphating of Iron-Based Metallic Parts in the Subtractive/Additive System

The technological procedure of crystalline lubricant phosphating of ferrous metallic parts in the **subtractive/additive** system includes several stages and phases shown in figure 3.3.

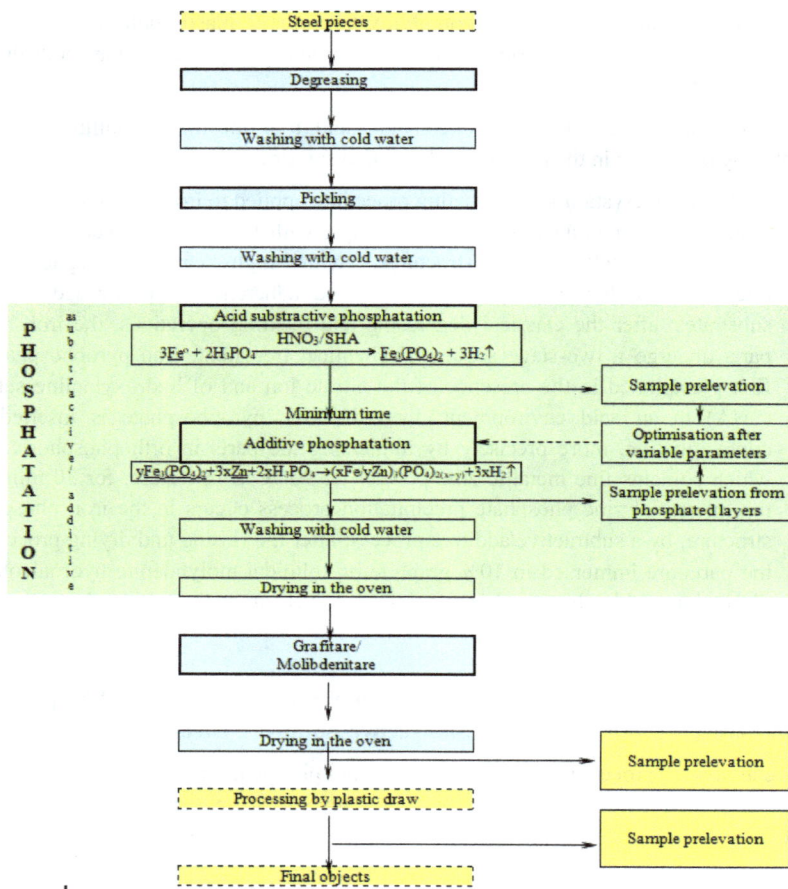

Fig. 3.3. *Technological flow of microcrystalline phosphating of ferrous metallic parts in the subtractive/additive system.*

According to the chart in figure 3.3, one may notice the following stages:

- steel blank creation; alkaline chemical degreasing; cold running water rinsing; acid chemical pickling; cold running water rinsing;

- acid subtractive phosphating according to reaction (3.1); control specimen sampling for subtractive phosphate coating testing to optimize variable parameters; additive phosphating according to reaction (3.2); cold water rinsing; phosphate-coated blanks dried in the oven at the set parameters; phosphate-coated control specimen sampling for additive phosphate film testing;

- graphite/molybdenite-coating; graphite/molybdenite-coated blank drying; graphite/molybdenite-coated control specimens sampled for testing to optimize the working parameters;

- plastic working of phosphate- and graphite/molybdenate-coated blanks; phosphate-coated control specimen sampling for phosphate film testing after plastic deformation; finite parts.

3.5   Chemical Composition and Work System Parameters Applying to Crystalline Chemical Phosphating in the Additive/Subtractive System

For crystalline chemical phosphating in the **additive/subtractive** system, the chemical composition of the solutions and the work system parameters applying to alkaline chemical degreasing, acid chemical pickling, additivation by deposition, subtractive phosphating and graphite/molybdenite-coating are described for subtractive phosphating based on both zinc cations and metallic zinc.

In case of additivation by initial deposition and of subtractive phosphating, there is a set of variable parameters used to optimize the crystalline chemical phosphating process in the **additive/subtractive** system.

3.5.1   Systems based on zinc cations

The procedure relies on phosphating in an acid environment by an **additive/subtractive** mechanism, in the presence of the Cu, Zn and Mn cations, stable in the oxidation state (II), which, after precipitation, become inert in relation to the Fe(0) in the substrate.

These cations are only susceptible to acid-base processes, which lead to the creation of thin porous films by orthophosphate coprecipitation, even in the presence of poorly hydrated oxidic or saline stains containing Fe(II, III), which are solubilized by slight surface oxidation, in the presence of permanganate anion.

The film thus formed is then dipped in graphite or molybdenite colloidal hydroalcoholic dispersions, which have undergone steric and electrostatic stabilization, in the presence of the NH_4OH-NH_4Cl buffer system, which ensures an optimum pH of 8.5...9.00.

The chemical composition and work system parameters used for alkaline chemical degreasing and for acid chemical pickling of steel parts by immersion in static tank are shown in table 3.3. and 3.4.

Table 3.3. *Chemical composition and parameters for alkaline chemical degreasing.*

Compound	Concentration, [g/L]
NaOH	40
Na_2CO_3	30
$Na_3PO_4 \cdot 10H_2O$	30
$Na_2SiO_3 \cdot 9H_2O$	5
Detergent (tensioactive)	3...10
Parameters	**Value**
Temperature, [°C]	80...90
pH	11...12
Time, [min]	10

Table 3.4. *Chemical composition and parameters for acid chemical pickling.*

Compound	Concentration, [g/L]
HCl ($\rho=1,19g/cm^3$)	150
Hexametilentetramine, $C_6H_{12}N_4$	0,45
$Na_2SO_4 \cdot 10H_2O$	0,15
Parameters	**Value**
Temperature, [°C]	20...25
pH	1...2
Time, [min]	max. 30

The chemical composition and work system parameters used for additivation by the initial immersion of steel parts in static tank are shown in figure 3.4.

Chemical compounds	Concentration	
$CuSO_4 \cdot 5H_2O$	0,5M	
Natural tanin (Quebracho or mimosa)	10, 15, 20, 25, 30 g/l	Variable parameter
H_2SO_4	câteva picături	

$$Fe^0 + CuSO_4 \rightarrow \underline{Cu^0} + FeSO_4$$

Work parameters	Value	
pH	3,5...4	
Temperature, [°C]	20, 40, 60, 80	Variable parameter
Timpe, [min]	1, 2, 3, 4, 5	

ADDITIVATION BY CEMENTATION

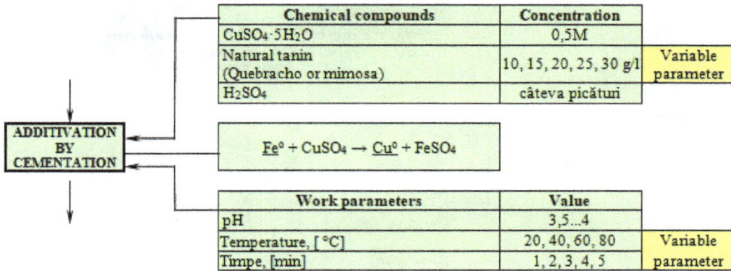

Fig. 3.4. Chemical composition and work system parameters for copper deposition.

The chemical composition and work system parameters used for subtractive phosphating based on zinc cations of the steel parts immersed in the static tank are shown in figure 3.5.

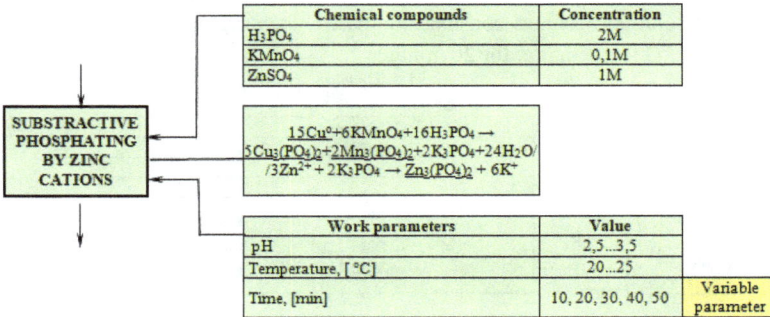

Chemical compounds	Concentration
H_3PO_4	2M
$KMnO_4$	0,1M
$ZnSO_4$	1M

$$15Cu^0 + 6KMnO_4 + 16H_3PO_4 \rightarrow$$
$$5Cu_3(PO_4)_2 + 2Mn_3(PO_4)_2 + 2K_3PO_4 + 24H_2O/$$
$$/3Zn^{2+} + 2K_3PO_4 \rightarrow Zn_3(PO_4)_2 + 6K^+$$

SUBSTRACTIVE PHOSPHATING BY ZINC CATIONS

Work parameters	Value	
pH	2,5...3,5	
Temperature, [°C]	20...25	
Time, [min]	10, 20, 30, 40, 50	Variable parameter

Fig. 3.5. Chemical composition and work system parameters used for subtractive phosphating based on zinc cations.

The chemical composition and work system parameters used for graphite/molybdenite-coating of steel parts coated with phosphate in the additive/subtractive system and immersed in the static tank are shown in figure 3.6.

GRAPHITE/ MOLIBDEN DEPOSITION	The objects are immers in hydro-alcoholic solutions with 10%graphite or coloidal molibden, stabilised in presence of an tensioactive agent (sodiul arilsulphonate etc.) and a buffer system made of NH₄OH-NH₄Cl, with 1,2% NH₄OH and 0,5% NH₄Cl.

Work parameters	Value
Temperature, [°C]	20...25
pH	8,5...9
Time, [min]	3...5

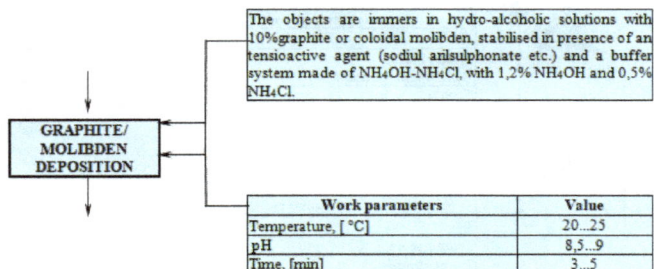

Fig. 3.6. Chemical composition and work system parameters for graphite/molybdenite-coating.

3.5.2 Systems based on metallic zinc

After the thin porous copper phosphate-based films have been deposited, the part is immersed for 30 minutes in an aqueous solution of 2M orthophosphoric acid, which contains 50 g of fine metallic zinc powder dispersed in 1 liter of solution; this is followed by successive rinsing operations by 10 minute immersion in distilled water, until the pH of the rinsing solution becomes constant; the part is then dried in a thermally regulated oven, at 110 ± 5 °C temperature, for 20 minutes.

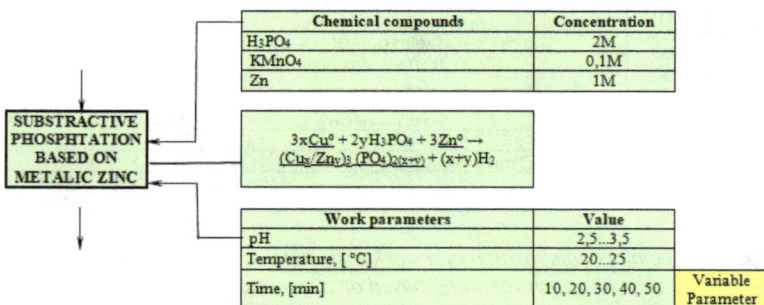

Chemical compounds	Concentration
H_3PO_4	2M
$KMnO_4$	0,1M
Zn	1M

SUBSTRACTIVE PHOSPHTATION BASED ON METALIC ZINC

$$3x\underline{Cu}^0 + 2yH_3PO_4 + 3\underline{Zn}^0 \rightarrow (Cu_x/Zn_y)_3 \,(PO_4)_{2(x+y)} + (x+y)H_2$$

Work parameters	Value	
pH	2,5...3,5	
Temperature, [°C]	20...25	
Time, [min]	10, 20, 30, 40, 50	Variable Parameter

Fig. 3.7. Chemical composition and work system parameters for subtractive phosphating based on metallic zinc.

The pickling, degreasing, copper additivation and graphite/molybdenite-coating operation are done with the solutions and parameters described previously. Subtractive phosphating is changed, and the change consists of replacing the zinc cations in the zinc sulfate with metallic zinc powder. The chemical composition and work system parameters for subtractive phosphating based on metallic zinc are shown in figure 3.7.

3.6 Chemical Composition and Work System Parameters Applying to Crystalline Chemical Phosphating in the Subtractive/Additive System

Degreasing, pickling and graphite/molybdenite-coating are done according to previously presented parameters. Figures 3.8 and 3.9 show the chemical compositions and working parameters for acid subtractive phosphating and additive phosphating, respectively.

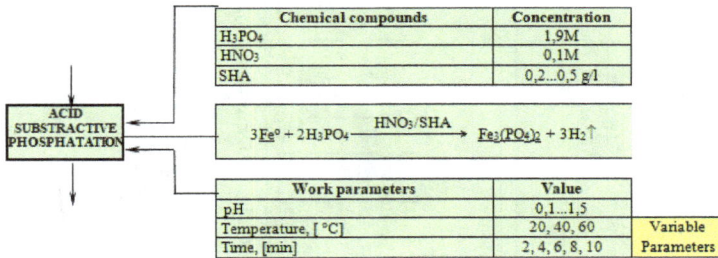

Fig. 3.8. *Chemical composition and work system parameters for acid subtractive phosphating.*

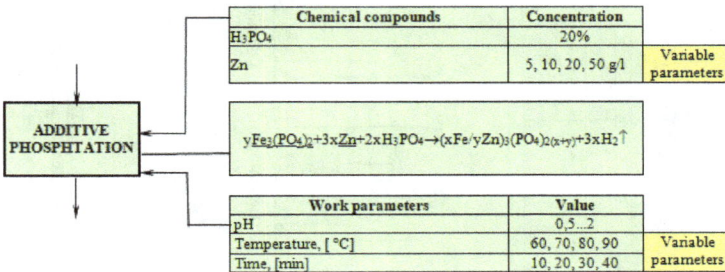

Fig. 3.9. *Chemical composition and work system parameters for additive phosphating.*

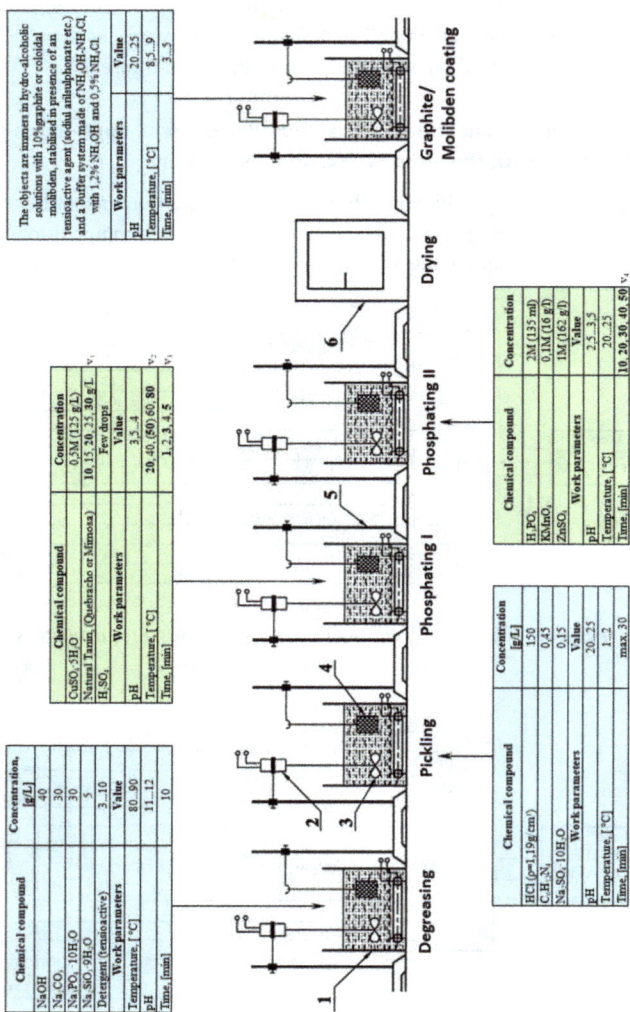

Fig. 3.10. *Laboratory equipment used for crystalline chemical phosphating in the additive-subtractive system based on zinc cations: 1- thermostat tank, 2 – mechanical stirrer, 3 – stirrer, 4 – part basket, 5 – universal holder, 6 – oven.*

3.7 Designing Laboratory Equipment for Crystalline Chemical Phosphating in the Additive/Subtractive and Subtractive/Additive Systems

Considering the work system parameters and their variations, especially temperature, thermostat tanks will be used for degreasing, pickling, copper deposition, subtractive phosphating, additive phosphating, graphite/molybdenite-coating, with the following technical characteristics: temperature range: room temperature up to 99.9 °C; temperature resolution 0.1°C; stability +/-0.15 °C, homogeneity +/-1 °C; exterior and interior structure of stainless steel; digital temperature display; heating elements of stainless steel inside the thermostat tank; hydraulic safety thermostat with overheating protection and low solution level; programmable timer between 0 and 99.5 hours; capacity, 5L.

Mechanical stirrers will be used to stir the solutions: 400 Ncm torque at 40 rpm; speed range, 40 – 2,000 rpm; digital display for reproducible settings. Also, an oven with the 5 °C – 300 °C temperature range will be used for drying. Figure 3.10 shows an example of phosphating laboratory equipment for the additive/subtractive system based on zinc cations. The other pieces of equipment used for the additive/subtractive system based on metallic zinc and for the subtractive/additive system are similar, except that the solutions used are modified accordingly.

Chapter 4

Thin Phosphate Film Characterization Methods and Techniques

4.1 Introduction

When referring to material investigation, the studies designed to determine their physical-structural and chemical properties are very important.

In order to determine the characteristics of thin films, we should first know what the right methods and techniques are. Therefore, it is necessary to know both the characteristics that require studying, and the scientific investigation methods, as well as the experimental data interpretation approaches. The approaches currently preferred consist of coassisted or corroborated interdisciplinary techniques, which allow good experimental data resolution and interpretation.

The thin film structure and morphology study methods rely on electromagnetic radiation interaction with matter. This interaction comprises a both elastic and non-elastic scattering. Depending on how these scattered electromagnetic radiations are detected, we speak of different study methods (diffraction methods, spectroscopic methods, etc).

The most common methods used for thin film study may be classified as follows:

- *microscopic methods* (optical microscopy, transmission electron microscopy, scanning electron microscopy, field ion microscopy, thermionic emission microscopy, atomic force microscopy, optical profilometry, etc.);

- X-ray, electron, neutron and gamma radiation diffraction methods;

- *spectrometric emission and absorption methods* in the visible, ultraviolet, infrared (FTIR), X-ray and gamma radiation fields;

- *radiospectrometric methods*, nuclear magnetic resonance (NMR), electron spin resonance (ESR), ferromagnetic resonance (FMR), etc.

For our research, we used optical and scanning electron microscopy (surface morphology), optical 3D profilometry (surface profile, roughness, size and distribution of surface structures), EDX (Energy-dispersive X-ray spectroscopy – elemental composition), XRD (X-ray Diffraction – composition, elementary cell and its parameters), tribology (friction coefficient determination), corrosion tests (corrosion resistance in saline environment), microFTIR analysis (chemical composition) and cold plastic working tests.

4.2   Optical Microscopy

4.2.1   Short History

1564-1642 Galileo Galilei, Italian astronomer, mathematician and physician, professor at the University of Pisa and Padova, is considered the founder of exact sciences and modern scientific methods. In 1610, he joined the precursors of modern compound microscope.

1580-1656, Francesco Fontana, Italian astronomer, was probably the first scientist who replaced the initial concave eyepiece of the microscope with a convex one.

1645, the name microscope was suggested by Demiscianus, member of the Lincei Academy, and

1597-1660, the names lens and eyepiece are the contribution of Schyrlaus Rheita, the inventor of terrestrial telescope.

1675 Robert Hooke (1635-1703) created the first compound microscope having a practical interest, thus becoming the pioneer of high precision microscopic observations. All these observations are described in his book, which has become a reference in the history of microscope, "Micrograpfia or some philosophical description of minute bodies", London, 1665.

4.2.2   Operating Principle

Optical microscopes are used to study the microstructure of a material. They may be used for transmission, when the specimens are thin and transparent, and for reflectance, in case of opaque materials. The contrasts in the image produced result from the reflection differences of the various areas of the microstructure [*Baciu, 1996; Briggs, 1990; Walls, 1989*].

Optical microscopes may normally magnify by 25 to 1000 times.

Optical microscopes may have a series of filters, namely:

- bright field, which is the most common metallographic analysis technique. The incident light is reflected back through the lens, thus achieving a bright background for flat surfaces and slightly darker for the others (pores, edges, boundaries of the grains that were etched), since the light reflected by those surfaces falls at a different angle;

- dark field, which is extremely useful for analyzing surface artifacts, such as surface scratches, for studying grain structure. The path followed by the light is slightly different, as it comes from the source, exits through the lens, is reflected back by the surface of the specimen and returns through the lens and then the eyepiece;

- polarized light, which is useful especially for analyzing anisotropic metals, such as beryllium, titanium and zirconium;

- differential interference contrast (DIC) microscopy is used to enhance the topographical characteristics of a surface. The method uses a Normarski prism, which divides the light into two beams. Together with a polarizer, it increases the relief effects of the surface.

- Optical interferometry is a technique which provides accurate and detailed information about the material surface topography. The simplest interferometer is based on the interference between two light beams. One beam is focused on the specimen and the second of a flat reference surface. The two reflected beams are then recombined by the beam separator and cross the eyepiece together. Today's interferometers provide quantitative information about the topography of the three-dimensional surface of the analyzed material.

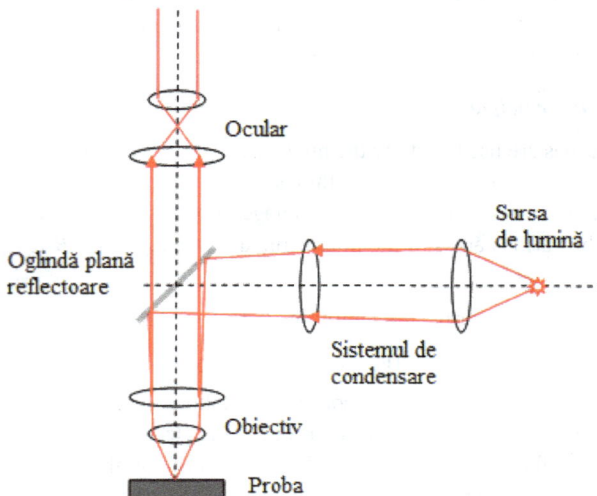

Fig. 4.1. *Sketch of an optical microscope.*

The light reflected by the object (surface of the specimen) crosses the lens, which creates an intermediate magnified reversed image. This image is then magnified by the eyepiece and forms a virtual image, visible with the naked eye, or a real image, projected on the computer screen, by the photo-video camera [*Marinescu, 1984; Suciu, 2008; Jantschi, 2004; Drake 1978*].

Optical microscopy has known a serious comeback during the last few years, due to the discovery of methods able to detect the diffraction limit of light. Unfortunately, all the methods also involve a significant increase of the manufacture and maintenance costs.

Here are some of these methods: SIL (Surface Immersion Lens), NAIL (Numerical Aperture Increasing Lens), NSOM (Near-field Scanning Optical Microscopy of Optoelectronic Devices), STED (Stimulated Emission Depletion fluorescence microscopy), SPM (Surface Plasmon Microscopy) [*Kasemo, 1997; Jantschi, 2004; Drake 1978; Kruger 2000; Lipson, 2010*].

4.2.3 Microscope Used for Our Tests

Our tests were performed using an optical Zeiss Imager Axio a1M microscope, with a 100 – 500 times magnification factor, which has dark field and bright field filters, as well as polarization filters (fig. 4.2) [www.zeiss.com].

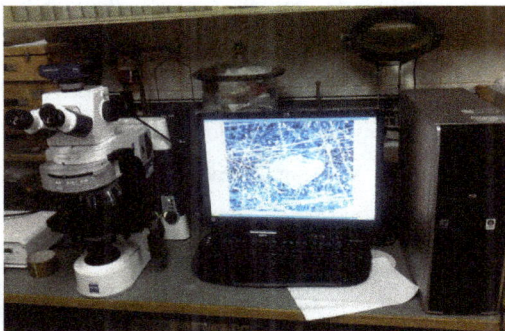

Fig. 4.2. *Optical microscope used for our tests.*

The reflection technique is used to study surface microstructure by optical microscopy. The contrasts in the image produced are the result of the reflection differences of the various microstructure areas [*Walls, 1989; Briggs, 1990; Baciu, 1996; Petraco, 2003;*

Suciu, 2008]. It is well-known that any optical microscope may în mod normal, poate mări între 20 până la maxim 1000X. Stereomicroscopy is very often used, as it allows the analysis of structure morphology and microtopography. This technique may magnify by 2.5 to 300 times.

The optical microscope used has a series of filters like Bright Field, Dark Field and polarized light.

The light reflected by the object (surface of the specimen) crosses the lens, which creates an intermediate magnified reversed image. This image is then magnified by the eyepiece and forms a virtual image, visible with the naked eye, or a real image, projected on the computer screen, by the photo-video camera [*Ilca, 1985; Scott, 1991; Kasemo, 1997; Petraco, 2003; Jantschi, 2004; Suciu, 2008*].

Before being analyzed, some specimens were either embedded in epoxy resin, in order to allow their cross-section examination, or they were used as such, the surfaces being flat.

4.3 Electronic Microscopy

4.3.1 Short History

Electronic microscope is an example of the manner in which some of the great physics discoveries made at the end of the 19th century were put into practice. These discoveries were the result of a serious research work on the behavior of electricity in vacuum, research that began as early as 1750. The following contributions are considered extremely important for today's electronic microscopes [*Ploaie, 1979*]:

- the discovery of cathode rays by Eugen Goldstein (1850-1930) in 1876;
- the discovery of X rays by Wilhelm Conrad Röntgen (1845-1923) in 1895;
- the demonstration of the existence of the electron by Joseph John Thompson (1856-1940) in 1897;
- the invention of the oscilloscope tube by Karl Ferdinand Braun (1850-1918) in 1895;
- the determination of the waveform character of electrons and their resemblance to photons by Louis de Broglie in 1923;
- the setting of the lens role of electric and magnetic fields and axial symmetry on electrons by Busch in 1926.

All these discoveries helped substantiating electronic optics the object of which is the study of the movement of electrons in vacuum under the influence of electric and magnetic fields. Electronic optics knowledge is the basis for the creation and development of electronic microscopes, particle accelerators, cathode ray tubes and of image analyzers and processors.

The first electronic microscope was built in 1931 by the German engineers Ernst Ruska and Max Knoll. It relied on the ideas and discoveries of the French physicist Louis de Broglie. Although primitive and unsuited for practical uses, the device was capable of magnifying objects by four hundred times.

Reinhold Rudenberg, the research manager of Siemens Company, had the electronic microscope patented in 1931, although Siemens did not carry out research in the field of electronic microscopes at that time. In 1937, Siemens started to finance Ruska and Bodo von Borries who developed an electronic microscope. Siemens also hired Ruska's brother, Helmut, to work for the applications, in particular cu biological specimens [*Ruska, 1986; Kruger, 2000*].

During the same decade, Manfred von Ardenne invented the scanning electron microscope and all-purpose electronic microscope [*von Ardene, 1940*].

Siemens started the series production of the transmission electron microscope in 1939; before that, the first electronic microscope with practical use had been built at Toronto University in 1938, by Eli Franklin Burton and his students Cecil Hall, James Hillier and Albert Prebus.

Although modern electronic microscopes may magnify objects by up to two million times, they are all based on Ruska's prototype.

4.3.2   Design and Operating Principle

In all types of electronic microscopes, the electrons generated by an electron gun and subjected to gun potentials of the user's choice are focused and dispersed to form an image, as they cross electrostatic or electromagnetic fields, depending on the microscope structure type. In such fields, the electrons may be determined to behave in an analogous way with the photons in the lens system of the photonic microscope. Since they behave as a lens towards an electron beam, electric or magnetic fields have been named electrostatic lens or electromagnetic lens, depending on their structure [*Ploaie, 1979; Bojin, 2005*].

Electrostatic lenses are metallic disks placed separately or in groups of three, which have a particular electric potential. These disks operate as axial symmetry electrodes. *Electromagnetic lenses* include a slenoid the coils of which are placed on a highly pure iron core with great permeability to the magnetic field. An axial cylindrical hole is drilled in the center of the iron core, which is 1...3 cm in diameter, or sometimes larger, depending on the type of microscope and on the role played by the lens (condenser, lens, etc.).

The magnetic field is concentrated in an inner slot of the coil and has a rotational symmetry with a symmetry axis which corresponds to the lens axis.

There are two types of *electromagnetic coils: long and short*, the latter having a very high number of turns. By changing the field current in the short coils, the magnetic field may be concentrated in the inner slot, which allows both decreasing the focal distance of the lens, and achieving high magnification values.

Polar parts are the most valuable components of an electron microscope and they are trade secrets of vital importance for microscope manufacturers. Polar parts are shaped like cone or cylinder frustums and they have a 1…3 cm in diameter axial hole drilled in the middle, which is crossed by the electron beam. Thus, the role of these parts is to attain the highest concentration of the lines of force of the magnetic field.

4.3.3   *Classification of Electronic Microscopes*

Existing electron microscopes are classified according to their structure and destination. Specialists are familiar with their classification and acronyms of their names in English are used in regular language [*Flewit, 1994; Ilca, 1985; Mitelea, 1987; Geru, 1985; Bunea, 1995; Gheorghiu-Dobre, 1998; Callister, 1985, Desy, 1968; Faya, 2005*].

Electron microscopes are therefore classified as follows:

- ➢ *TEM – Transmission Electron Microscope*, used for ultrastructural research;
- ➢ *SEM – Scanning Electron Microscope*, used in surface morphology research with the help of secondary or reflected electrons;
- ➢ *STEM – Scanning Transmission Electron Microscope*, which allow the structural study of specimens by transmission and of surfaces by SEM;
- ➢ *TEAM – Transmission Electron Analytical Microscope*, with simultaneous structural and analytical applications;
- ➢ *Systemic Electron Microscopes*, complex devices with system functionality which allow multiple simultaneous research;
- ➢ *PEM – Photoelectron Emission Microscope*, similar to SEM;
- ➢ *EPI – Electron Probe Instrument*, used for elementary analysis and surface study;
- ➢ *FIM – Field Ion Microscope*, which allow direct viewing of the manner in which atoms are oriented in a material.

4.3.4   Scanning Electron Microscope – SEM

Scanning electron microscopes allow the analysis of specimens 5 to 40 mm thick, with irregular surfaces, and they provide 3D images of the analyzed objects [*Ploaie, 1979, Callister, 1985, Desy, 1968*].

The image is formed with the help of secondary electrons (SE) or backscattered electrons (BSE), which occur after the specimens have been bombarded by the primary electron beam.

The electron beam produced by the electron gun is decreased down to its minimum by 2 or 3 electromagnetic lenses, the purpose being an extremely narrow beam with a diameter below 100Å, which is projected on the specimen.

By using two deflection coils located inside the last electromagnetic lens, activated by a current produced by a scanning generator, the primary electron beam thus focused is determined to **zigzag (raster)** over the specimen, thus "scanning" its surface. Therefore, the probing beam is found in different points on the surface of the specimen at different moments in time. Secondary electrons are captured by a detector, and the current formed by them is amplified and sent to a cathode ray tube, where it is used to modulate the intensity of the beam of this tube

The scanning generator, which sends a saw tooth-shaped current in the deflection coils of the microscope in order induce the raster movement of the preparation beam, simultaneously sends a current with the same shape and intensity in the deflection coil of the cathode ray tube, thus causing its electron beam to perform a similar raster movement.

The brightness of an image point on the monitor screen depends on the number of secondary electrons leaving the specimen from that point.

In all SEM microscopes, the resulting image of the analyzed object is sequential in time, as the same principles used in television apply here.

Scanning electron microscopes include 3 distinct interconnected functional units, namely: *electronooptic system; vaccum system and operating and display system.*

Figure 4.3. shows the construction drawing of a scanning electron microscope.

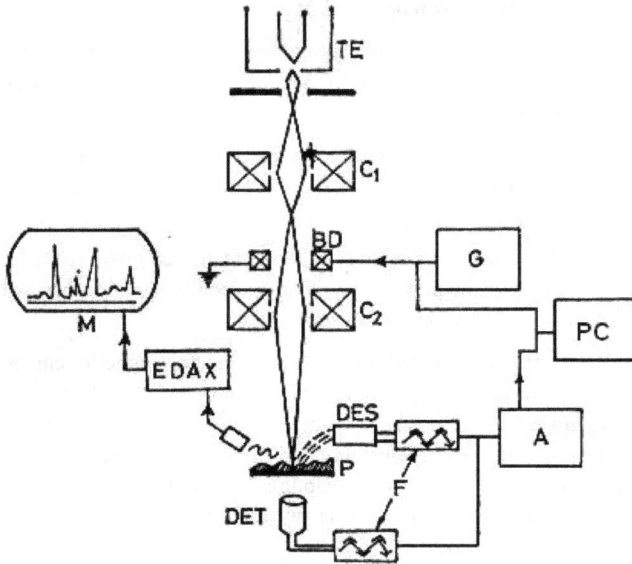

Fig. 4.3. *Construction drawing of a scanning electron microscope [Callister, 1985, Desy, 1968]: TE – electron gun; C_1 and C_2 – condenser lens; BD – deflection (scanning) coil; P – specimen to analyze; DES – secondary electron detector; DET – transmitted electron detectors; EDAX – X-ray detector; M – monitor; F – photomultiplier; G – scanning generator; A – amplifier; PC – display screen.*

a. The electronooptic system consists of the microscope column, specimen chamber and detector system.

The microscope column does not exceed 80 cm in height and is usually placed on the same table as the operating system and the display.

The *electron gun* is located at the top of the column and it generally uses *tungsten thermocatode* (filament) triodes.

The gun potential applied does not exceed 60 KV and it is applied at different increments starting at 1000V, depending on the specimen. The diameter of the accelerated electron beam when it leaves the Wehnelt cylinder is 250,000...500,000Å. In order for it to be used for probing, it is narrowed down to a 100Å diameter or even smaller, and lowered at specimen level. In some microscopes the decrease is achieved by means of two or three condenser lenses. These electromagnetic lenses are the main constituent of the microscope column.

The last lens is the most important and it is called *objective lens*, although its role is the final focusing of the beam on the specimen. In its middle it includes a deflection or scanning beam system and a stigmator to correct lens astigmatism.

The specimen chamber and the place where the specimen signal detectors are attached are also at the bottom of the column.

The detector system is the most important part of a microscope, which allows one or more device operating modes. The basic default system, which all standard microscopes have, includes a *secondary electron* and *backscattered electron* detector.

Each detector is connected to an electron unit mounted on the control console. The control units allow easy switching between the captured signals, provided the device is equipped with all types of detectors.

The electron detector is made up of a collector, a scintillator and a photomultiplier.

The general detector system is able to provide information about: geometric relief and topography, distribution potential and composition.

b. Vacuum system. The lower part of the specimen chamber enters into contact with a vacuum system, which generally includes one or more vacuum pumps. The vacuum may be: high ($<10^{-5}$ torr), medium (10^{-5} torr) and low ($>10^{-5}$ torr).

c. Operating system and display. This system makes up the control console and nowadays a computer is used.

The microscope may be controlled by the computer, which can change parameters and record images.

The following parameters may be changed: gun potential, scanning rate, magnification increment, focusing distance, astigmatism, beam size, etc.

4.3.5   *Microscope Used for our Analyses*

In our research, we used a SEM microscope, VEGA II LSH model, manufactured by TESCAN Czech Republic, coupled at an EDX detector type QUANTAX QX2, manufacturer - BRUKER/ROENTEC Germany.

The fully computer controlled microscope has an electron gun with tungsten filament, which may achieve a 3nm resolution at 30KV, with a magnification power ranging

Materials Research Forum LLC
doi: http://dx.doi.org/10.21741/9781945291913

between 30 and 1,000,000 X in the resolution mode, a gun potential of 200 V to 30 kV, and a scanning rate of 200 ns to 10 ms per pixel. The working pressure is below $1x10^{-2}$ Pa.

The microscope also has an EDX Quantax QX2 detector described in the next subchapter.

Figure 4.4. shows the TESCAN VEGA II LSH electron microscope.

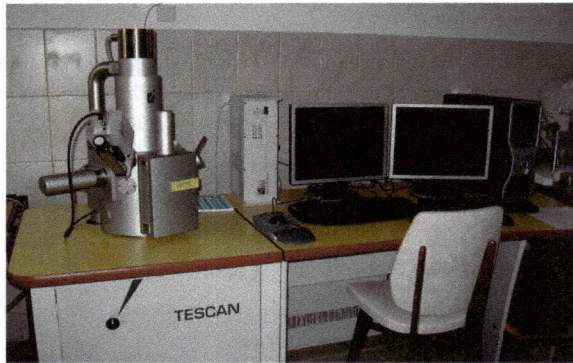

Fig. 4.4. *VEGA II LSH electron microscope connected to an EDX X-ray detector.*

The SEM VEGA II LSH electron microscope, connected to a EDX of the QUANTAX QX2 type, is able to perform a ***microscopic analysis*** and identify the microstructure, phases and constituents by using SE (Secondary Electrons – 3D image of the surface) and BSE (Back Scattered Electrons –2D image of the surface, better contrast of different phases) detectors, magnification increment ranging between 50X and 100,000X (1,000,000X) [27];

4.4   EDX Analysis

X rays or **Röntgen rays** are electromagnetic ionizing radiations, with small wavelengths, ranging from 0.1 to 100 Å [*Flewit, 1994; Mitelea, 1987; Faya, 2005; Preston, 1991; www.tescan.com*].

4.4.1 Short History

1895 - discovery of rays by Roentgen

1912 - Friedrich and Knipping confirm that X rays may be defracted by crystals

1913 - Braggs creates the first X-ray spectrum for Pt by using a NaCl crystal

1913 - Mosely discovers the systemic variation of wavelengths specific to X rays of different elements (the wavelength is inversely proportional to Z^2)

1922 - Hadding uses X-ray spectra for mineral analysis

1923 - von Hevesy discovers Hf after having noticed that something was missing in Z=72

1920-1930 - transmission electron microscopes are created in Germany and the first practical demonstration takes place in 1932 - Ernst Ruska (who won the Nobel prize in 1986), the prototype being built by Siemens & Halske Co.

1930-1940 - scanning coils are included in TEM , which thus become STEM (the image is produced by the secondary electrons emitted by the specimen)

1940 - RCA sells the first TEM outside Germany

1942 - the first use of a SEM to analyze the thickness of certain layers in RCA laboratories

1949 - Castaing builds the first microscope used for microchemical analysis (with a spectrometer with wavelength dispersion focus crystal = WDS)

1965 - starts the series SEM production

1968 - the first EDX solid state detectors are created

4.4.2 Operating Principle

Energy-dispersive X-ray spectroscopy, also known as EDS, EDX or EDAX, is a technique used to identify the elemental composition on a small area. During the analysis, the specimen is subjected to an electron wave released by a scanning electron microscope (SEM). These electrons hit the specimen electrons and thus push some of them off the orbits. The places left vacant are occupied by higher energy electrons, which emit X-rays during the process. The elemental composition may be determined by analyzing the emitted X-rays. EDX analysis is an important tool in elemental constituent analysis. The most suitable materials to analyze are: metals, alloys, ceramics and minerals [*Bojin, 2005; Flewit, 1994; Faya, 2005*].

The X-rays emitted by the bombarded specimen are captured by a detector (cooled by liquid nitrogen – old models and with peltier module – new models) and analyzed by a computer.

The resulting spectrum will display a real-time histogram of the number of X-rays detected per channel (which is variable, ranging between 1 and 10 electrons volts/channel) as compared to the energy dispersed in KeV.

In practice, EDX is most frequently used for elemental analysis, in order to identify the elements and their relative presence. Depending on the analysis aimed at, a quantitative analysis is also possible further to spectra assessment, by comparisons with the reference standards in the database and by using various algorithms. All these have been recently included in the analysis software of the microscope.

4.4.3 Detector Used for the Analysis

In our research we used an QUANTAX QX2 EDX detector, manufactured by BRUKER/ROENTEC Germany, connected to a scanning electron microscope SEM, VEGA II LSH model, manufactured by TESCAN Czech Republic (Fig. 4.4.).

Quantax QX2 is an EDX detector used for qualitative and quantitative micro-analysis in industry, research and education, which performs quantitative measurements without using specific calibration standards. Its active area is 10 mm^2, and it is able to analyze all elements heavier than carbon, ground or irregular specimens, thin films or particles, with a resolution below 1.33 eV (MnKα, 1000 cps). Quantax QX2 uses a 3rd generation detector, Xflash, a detector that does not need liquid nitrogen cooling and that is about 10 times faster than conventional detectors Si(Li).

Thus, the QUANTAX QX2 EDX detector, connected to the VEGA II LSH SEM microscope, is able to perform the following types of analyses:

- *chemical qualitative and quantitative analysis of the elements* (all with the atomic radius greater than that of Carbon, except for rare gases, lanthanides and actinides) on the surface of the specimen;
- *chemical qualitative and quantitative analysis* of certain constituents in certain places on the surface of the specimen;
- *chemical analysis along a vector*, on the surface of the specimen;
- *element mapping on the analyzed surface*, in order to identify the segregations, separations, distinct phases and evenness and dispersion of elements.

4.5 XRD Analysis

4.5.1 Operating Principle

Diffraction methods are methods that analyze elastically dispersed electromagnetic radiation. These methods include both X-ray diffraction, and electron and neutron diffraction, respectively.

X-rays may be produced either by accelerating or decelerating the loaded particles, called bremsstrahlung, or as an effect of radial acceleration of an electron or positron beam having a circular path, synchrotron radiation [*Moore, 1997*].

X bremsstrahlung and specific X radiation are obtained by using Röntgen tubes. High energy electron braking in the anode material, leads to X bremsstrahlung with broad photon energy spectrum (continuous spectrum, white radiation). The minimal wavelength is independent of the nature of the anode, as it only depends on the gun potential. In addition to the continuous spectrum of the bremsstrahlung, some peaks may also be noted at well-defined wavelengths. The positions of these peaks depend on the nature of the anode and they form the specific X-ray spectrum [*Cullity, 1978; Blakemore, 1969*].

A series of approximations are used to simplify X-ray diffraction phenomenon treatment, namely:

- the vibrations of the crystal lattice are not considered, meaning that the ideal crystal is considered instead, with the atoms in their position of equilibrium;
- the incident rays on the crystal are thought to form a perfectly parallel beam;
- the absorption phenomenon (further to which radiation intensity would decrease) is overlooked;
- diffracted radiations (by reflection or transmission) are thought not to interact with the incident radiations or with the atoms found on their path.

Figure 4.5 shows the geometry of an X-ray scattering experiment. A well-collimated X-ray beam is directed on a specimen, where it is scattered. When the scattering is elastic, we speak of X-ray diffraction.

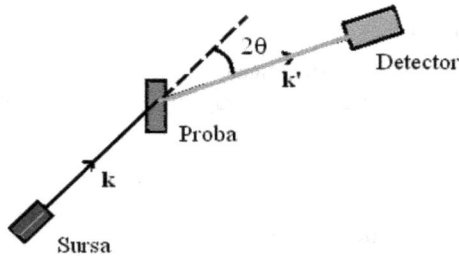

Fig. 4.5. *Elastic X-ray scattering. The wave vectors of the incident and emergent radiation are k and k', whereas the scattering angle is 2θ.*

The detector measures the intensity of the scattered wave, which is the result of the interference phenomenon, depending on the 2θ scattering angle. If the specimen is crystalline, the scattering angles corresponding to the diffraction peaks (interference of two coherent waves) are used to determine its structure.

In the vast majority of X-ray diffraction experiments, the incident wave may be considered a monochromatic flat wave.

Similar waves reach the detector from all the specimen particles, the resulting wave being their sum.

Since the particle-detector distance is almost the same for all the specimen particles, its value may be considered constant.

The readings of the detector are proportional to the intensity of the scattered wave.

The multiple radiation scatterings are negligible from the viewpoint of X-ray diffraction, for which reason the X-ray is thought to be scattered only once on each particle. In order to find the wave scattered on the specimen we should sum up the waves scattered by each particle in the specimen [*Warren, 1990; Pecharsky, 2008*].

To conclude with, we may say that, further to elastic X-ray scattering on the surface of a crystal, the detector will capture peak intensity waves provided the following conditions are met:

1) Bragg's condition (Laue condition)

2) the structure factor is non-zero.

The diffraction peak intensity is also influenced by certain experimental factors, more precisely by the mechanical or heat processing of the specimen.

4.5.2   *Characteristics of a Diffraction Image*

The image in X-ray diffraction of a material depends on the diffraction method and on the structural characteristics of the material. In case of the diffractometric method, the diffraction image may have the following characteristics:

- ➢ the diffractogram is the image in X-ray diffraction of a crystalline phase or of a mixture of crystalline phases;

- ➢ the diffractogram consists of a sequence of diffraction peaks, with the intensity of the diffracted X-ray on the ordinate, measured in impulses/second, and the 2θ angle on the abscissa, where θ is Bragg's angle, measured in degrees;

- ➢ for a full diffractogram, the 2θ angle ranges from $0°$ to $180°$, but in practice, depending on the material analyzed and on the information sought, the measuring range may be more or less limited.

The λ wave length of the X-ray used depends on the material that the anode of the X-ray tube is made of. X-ray tubes with Cu, Cr, Mo, Co, etc. anode are usually used, and the X-ray emitted by the tube is made monochrome by means of selective absorption filters or monochroming crystals, which allow only the use of the K_α component in the X-ray spectrum. The places of the diffraction peaks in the diffractogram change when another wavelength is used.

The aspect of the diffractogram is influenced by a series of factors. Whereas the diffraction peaks of the perfectly crystalline specimens are displayed as lines, the peaks of the flawed specimens are wider, and those of amorphous specimens are superimposed [*Cullity, 1978; Blakemore, 1969*].

Each diffraction peak corresponds to a certain crystalline plane (or family of planes) characterized by Miller indices. Depending on the symmetry group related to the crystal, the number of peaks, their place and splitting differ.

The research carried out using modern diffractometers provides information about: the type of crystalline structure, qualitative analysis of the phases, presence of certain crystallographic textures, analysis of certain internal tensions, etc.

4.5.3   Devise Used in our Research

Figure 4.6 shows the Bruker XRD used in our research [*www.bruker-axs.com*].

Fig. 4.6. *Bruker D8 FOCUS XRD.*

4.6   Optic Profilometry

4.6.1   Short History

Optic profilometry is an important technology used to study surface topography. Its main advantages are: short acquisition time (a few seconds for an image) and relatively broad visual field (which may reach several square millimeters).

The profilometer is a measuring tool used to analyze the profile of the surface, in order to quantify its roughness and more [*Bennet, 1989; Stout, 2000*].

The optic profilometer uses the optic interferometry technique. Its first important application was stellar interferometry.

In 1868 Hippolyte Fizeau first underlined the basic concept of stellar interferometry and the manner in which the interferences among the stars may be used to measure their size. Brown &Twiss defined intensity interferometry in 1956, by correlating intensities with magnetic fields.

The Michelson interferometer is the most common optic interferometry configuration and it was invented by Albert Abraham Michelson. Together with Edward Morley, he used an interferometer for the famous Michelson-Morley experiment conducted in 1887, in order to show light speed constant through several inertial frames, which made it unnecessary to use luminiferous ether to stop the light.

4.6.2   Operating Principle

Non-Contact Profilometers

The optic profilometer is a non-contact tool which provides similar information to basic stilus profilometers. There are many techniques involved here, such as: laser triangulation, confocal microscopy, interferometry and digital halography.

Most optic profilometers are interference microscopes used to measure height-roughness variations, being very accurate [Bennet, 1989, 2008; Dufour 2010].

Advantages of optic profilometers

Resolution: vertical at nanometric level, the lateral resolution is limited by the wavelength of light

Speed: since it is non-contact, the speed is influenced by the light reflected by the surface and by the acquisition speed of the system.

Reliability: very high, due to the fact that it does not touch the surface. Its maintenance costs are also much lower.

Size of spot: up to several tens of nanometers.

Optic profilometry is a fast nondestructive and contactless technique used to analyze surfaces.

A profilometer is usually a microscope in which the light is split in two: the first beam is directed towards the tested surface, whereas the second is directed towards the reference mirror. The reflections of the two surfaces are combined and projected onto a detector. The interference contains information about the surface (Fig. 4.7.). Optic profilometry uses the properties of light waves to compare the difference between the reference and the tested surface.

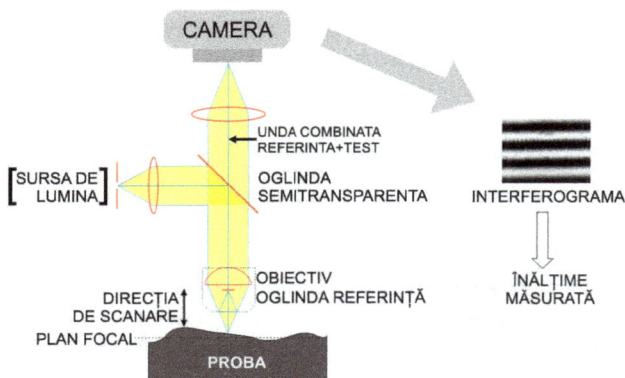

Fig. 4.7. Operation chart of an optic profilometer.

Since the flatness of the reference mirror is known and it is almost perfect, the differences are generated by the height variations on the surface of the specimen.

The interference beam is focused in a digital chamber, which sees the constructive interference areas as brighter, and the destructive interference areas as darker.

In the interferogram above, each transition from bright to dark is half of wavelength of the difference between the reference and the surface of the specimen.

Fig. 4.8. Example of interferogram.

When the wavelength is known, it is possible to calculate the height difference on the surface, in wave fractions. A 3D map of the surface may be drawn by using these height differences.

Materials Research Forum LLC
doi: http://dx.doi.org/10.21741/9781945291913

4.6.3 Device Used in our Research

We used a multi-sensor system called AltiSurf 500 (Fig. 4.9.).

It measures a few profiles, which when automatically put together allow building the morphology of the specimen. A piece of processing software analyzes parameters like: roughness, tribology, traces and sizes, and also topographic phenomena.

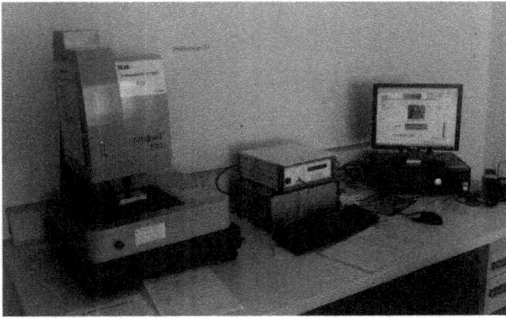

Fig. 4.9. *Profilometer used in our research – ALTISURF 500.*

4.6.4 Operating Principle

A white light spot scans the surface through a lens with axial chromatic aberration.

Instead of focusing in one point, the lens functions like a prism and separates the wavelengths, which represent one point on the analyzed surface. A perfect wavelength focus (the most intense after spatial filtering) generates the highest peak, which, converted by a CCD spectrometer (which measures the energy of the incident photons by means of a **"charge-coupled device"**), measures the height in that particular point.

4.6.5 Determination of the Surface Roughness

The roughness measurements may be performed on a rough area selected by the image. There are many roughness measurements [*Bennet, 1989, 2008; Dufour 2010*].

Root mean square roughness, **RMS**, is the standard deviation of the z value in a given surface. It is defined by the ratio:

$$RMS = \sqrt{\frac{\sum_{i=1}^{N}(z_i - z_{med})^2}{N}}$$

(7.1)

where Z_{med} is the mean value of $z(t)$ in the given surface, Z_i is the current $z(t)$ value and N is the number of points in the surface.

Mean roughness, Ra, is the arithmetical mean of the deviation from the middle of the plane and it is described by the ratio

$$Ra = \frac{\sum_{i=1}^{N} |Z_i - Z_{cp}|}{N} \qquad (7.2)$$

where Z_{cp} is the value of $z(t)$ in the middle of the plane, Z_i is the current $z(t)$ value and N is the number of points in the chosen surface.

The RMS and Ra formulae are to be found in the user's manual of the NanoScope software.

R_{max} is the height difference between the highest points and the lowest points in the surface, as compared to the medium plane.

The surface area is the 3D area of a particular surface. This value is the sum of the areas of all the triangles formed by three adjacent points.

The surface area difference is the 3D surface area increase percentage in comparison to the 2D surface area.

4.7 Corrosion Tests

Corrosion is defined as the degradation of a solid body under the unintentional chemical or electrochemical action of the environment. The detectable effects of corrosion are changes in the weight, alteration of the surface and weakening of the mechanical properties of the metallic part [*Nemtoi, 2007; Petrescu, 2009*].

Corrosion may be classified as follows:

- uniform corrosion, when the metal is corroded in a uniform manner, and its mechanical resistance decreases proportionally to thickness diminution;

- localized ("pitting") corrosion, when only points or parts of the metal is corroded and the resulting surface is grooved. This decreases the plastic deformation capacity, like iron in seawater;

- intercrystalline corrosion, when the metal is corroded in depth down to the grain boundary, without losing weight and sometimes without altering the surface. The mechanical properties suffer major changes, their breaking strength being very low, due to intergranular corrosion products.

From the viewpoint of its mechanism, corrosion may occur by dissolution, while in the dry state by oxidation or electrochemically.

4.7.1   Factors Influencing Resistance to Corrosion

The factors may be grouped in the following categories:

a) **Metallurgical factors** which refer to the material: nature of the alloy, degree of purity, its constitution and structure. They depend on the specificity of the technological process employed for the finite product. These factors may lead to heterogeneity as a consequence of the different resistance to corrosion for the same alloy.

b) **Chemical environmental factors**: nature of the corrosive environment, presence of impurities, concentration, pH, temperature, pressure, viscosity, dissolved oxygen content, presence of microorganisms.

c) **Use conditions factors**: shape of the part, status of the surface, assembly processes, mechanical stress, immersion conditions in the corrosive environment, etc.

If we consider all these factors, we may conclude that there are practically no metals or allows that cannot be corroded. A metallic material may be considered resistant to corrosion only when the nature of the corrosive environment and the conditions of use are known. Practically speaking, the cumulated effect of the mechanical stresses and of the action of the chemical environment leading to specific corrosion is very important.

4.7.2   Operating Principle

Corrosion tests were conducted on the created specimens, and the control part was a standard specimen made of soft DC01AM steel.

All the experiments were carried out at room temperature (25°C), using a 3-electrod electrochemical cell (fig. 4.10.b) connected to the Autolab PG STAT 302N (Metrohm Autolab) potentiostat shown in figure 1a, which also includes the Nova 1.6 software. The reference electrode, against which the potential values will be expressed, was made of saturated calomel (ECS), a flat platinum electrode was used as auxiliary electrode and the working electrodes were the specimens.

The cyclic voltammograms (CV) of the specimens were recorded in the -1.2÷1.8V range at a scanning speed v=100 mV/s, and they revealed point corrosion, as hysteresis specific to this type of corrosion occurred on back scanning [*Petrescu, 2009; Sutiman, 1999; Nemtoi, 2010*].

a   b

Fig. 4.10. *Autolab PG STAT 302N potentiostat (a) and electrochemical cell (b).*

In order to determine the corrosion parameters, linear voltammograms were drawn in the ±150mV range against the open circuit potential (OCP) of each specimen, determined before the drawing of the linear voltammogram. The OCP values were synthesized in a table, whereas the corrosion potential was calculated by means of the Nova 1.6 software, $E_{cor,calc}$ considers the possible changes that occur during LV drawing, yet it is not very different from OCP. When calculating the corrosion speed (expressed by the penetration speed), iron oxidation at $Fe(OH)_2$ was considered to take place as shown in literature [*Sanchez, 2006, 2007; Ningshen, 2009; Gordin, 2005*], by considering current density, j, defined by the Butler-Volmer ratio [*Armijo, 1967; Bockris, 1970*]:

$$j = j_{cor}\left[\exp\left(\frac{2,303(E - E_{cor})}{b_a}\right) - \exp\left(-\frac{2,303(E - E_{cor})}{b_c}\right)\right] \tag{8.1}$$

where b_a and b_c are the Tafel slopes:

$$b_a = \frac{RT}{\alpha nF} \qquad \text{and} \qquad b_c = \frac{RT}{(1-\alpha)nF} \tag{8.2}$$

Many researchers have noticed, by experimental means, that j varies approximately linearly with the applied potential (E), starting from approx. 50 ... 60 mV against the corrosion potential and only on a range of approx. 10 – 20 mV [*Jones, 1992*]. Stern and Geary [*1957*] simplified the Butler – Volmer equation for small surges against the E_{cor}. The simplified equation obtained may be written as follows:

$$R_p = \left(\frac{dE}{dj}\right)_{E_{cor}} = \frac{b_a \cdot b_c}{2,303 j_{cor}(b_a + b_c)} \quad (ohm \cdot cm^2) \tag{8.3}$$

After equation rearrangement, the following instantaneous corrosion current ratio results:

$$j_{cor} = \frac{b_a \cdot b_c}{2{,}303(b_a + b_c)R_p} \quad (mA/cm^2) \tag{8.4}$$

Thus, instantaneous corrosion current may be assessed directly from Evans diagram, based on the fact that the intersection of the Tafel lines provides the log j_{cor} value on the current density axis, and E_{cor} on the potential axis.

The polarization resistance may be calculated graphically:

$$R_p = \left(\frac{dE(mV)}{dI(mA)}\right)_{E_{cor}} = tg\alpha \quad (\Omega) \tag{8.5}$$

In order to calculate the instantaneous corrosion current, according to equation (4), the Tafel b_a and b_c constants are necessary, which are the slopes of the linear portions of the anodic and cathodic branches of the polarization curve in $E = f(\log I)$ coordinates:

$$b_a = \left(\frac{dE}{d\log I}\right)_{anodic} \quad (mV) \quad \text{and} \quad b_c = \left(\frac{dE}{d\log I}\right)_{catodic} \quad (mV) \tag{8.6}$$

The corrosion speed may be correlated with the corrosion current intensity or with the current density based on the general electrolysis law.

The surface corrosion rate was defined as the weight loss, in the time unit per unit of surface, expressed by the ratio:

$$V_s = \frac{\Delta m}{t \cdot S} \tag{8.7}$$

According to Faraday's 1st law:

$$\Delta m = K \cdot Q = K \cdot I_{cor} \cdot t = \frac{A \cdot I_{cor} \cdot t}{z \cdot F} \tag{8.8}$$

where Δm represents the specimen mass decrease during the t time, I_{cor} – corrosion current, $K = A/zF$ is the electrochemical equivalent of the corroded metal (A – atomic mass, z – number of electrons involved in the oxidation process, $F = 96485$ C/mol – Faraday's constant). By dividing ratio (8.8) by the $S \cdot t$ product:

$$\frac{\Delta m}{S \cdot t} = \frac{A \cdot I_{cor}}{zFS} \tag{8.9}$$

and considering the definition of the corrosion rate and the definition of the current density ($j = I/S$), we get:

$$V_s = \frac{A}{zF} j_{cor} \quad (g/cm^2 \cdot s) \tag{8.10}$$

The penetration speed, defined by the thickness of the metal layer, d, removed by corrosion in the unit of time and expressed by the following ratio:

$$v_p = \frac{d}{t} = \frac{v_s}{\rho} \qquad (8.11)$$

changes depending on the current density by replacing the surface corrosion rate expression in equation (10), when we get:

$$v_p = \frac{A}{zF\rho} j_{cor} \quad (cm/s) \qquad (8.12)$$

equations in which A/z is the $(1/z^*A)$ entity mole expressed in g/(mol), $F = 96485$ C/mol, ρ – corroded metal density (g/cm^3) and j – current density (A/cm^2). As the two speeds v_s and v_p are generally very low, more adequate units of measurement are used in practice, which rely on longer time intervals, so that the penetrate rate calculated in this paper by the Nova 1.6 software (shown in table 8.1), after having drawn the Evans diagrams, is as follows:

$$v_p = \frac{A}{z} \cdot \frac{j_{cor}}{\rho} \quad (mm/an) \qquad (8.13)$$

where $A=M_{Fe}, z=2, \rho= \rho_{Fe}$ and j_{cor} is calculated based on ratio (8.4).

The corrosion parameters were determined by using the Tafel lines in the Evans diagrams, as shown in Table 8.1.

Table 8.1. *Corrosion parameters of the specimens analyzed in aqueous NaCl solutions.*

Specimen	-OCP (mV)	b_a (mV/dec)	b_c (mV/dec)	$E_{cor,calc}$ (mV)	j_{cor} ($\mu A/cm^2$)	v_{cor} (mm/an)	R_p (Ω)
		Tafel slopes			Intensity		
Specimen code		$b_a = \dfrac{RT}{\alpha nF}$	$b_c = \dfrac{RT}{(1-\alpha)nF}$	Current voltage	$j_{cor} = \dfrac{b_a \cdot b_c}{2,303(b_a + b_c)R_p}$	Corrosion rate	Resistance

4.8 FTIR Spectroscopy

Infrared radiation (IR) is that part of the electromagnetic spectrum extending from the visible to the microwave areas, which is characterized by wavelengths of the 10^{-5} m order. Only the middle IR range is used to record IR spectra employed to determine the structure of the organic compounds, which contains wavelengths within the 2.5-25 μm range (most of the times, the characterization is done by using wave numbers within the 400-4000 cm^{-1} range).

IR radiation characterized by wave number below 100 cm^{-1} may be absorbed by the organic compound molecules and converted to molecular rotation energy. This absorption is quantified, which determines the recording of molecular rotation spectrum made up of discrete lines. IR radiation within the 10,000-100 cm^{-1} range may also be absorbed by the organic compound molecules, thus leading to changes in the molecular vibration states. Although this absorption is also quantified, the vibration spectra recorded by IR spectroscopy are made up of absorption bands, since each vibration energy change is accompanied by rotation energy changes.

An IR spectrum contains absorption bands due to the vibrations that occur simultaneously with the involvement of all the atoms in the structure of the molecules of the organic compound under survey (*normal vibrations*). The place of an absorption band formed by the vibration excitation of a particular functional group is well defined in the spectrum, and it varies within narrow limits with the ambiance of the functional group inside the molecule. An absorption band specific to the same functional group is found with almost the same value of the wave number in the IR spectrum of any molecule (*specific group vibrations*). This allows identifying the structural elements that make up a molecule, by assigning the specific absorption bands in the IR spectrum.

FTIR spectroscopy based on the two types of specific group vibrations (valence and deformation) allows gathering very precious information about the compounds formed as a result of surface phosphate precipitation processes.

It is a very modern technique coassisted by optical microscopy, which enables one to choose surface structures without specimen sampling, on very limited areas. Moreover, this technique may be corroborated with X-ray diffraction (XRD) and SEM-EDX data [*Bentley, 1968; Nakamoto, 1997; Coates, 2000; Salzer, 2009*].

In our case, the IR spectra were recorded by means of a FT-IR spectrophotometer connected to a HYPERION 1000 microscope, both manufactured by Bruker Optic, Germany (Fig. 4.11).

The FT-IR spectrophotometer is of the TENSOR 27 type, which is especially suitable for close IR measurements. The standard detector is DLaTGS, which covers the 7500 – 370 cm^{-1} spectral range and works at room temperature. Its usual resolution is 4 cm^{-1}, but it can reach 1 cm^{-1}. TENSOR 27 is equipped with a He – Ne laser emitting at 633 nm and a power of 1 mW, and it enjoys ROCKSOLID interferometer alignment. The signal/noise ratio of this device is very good. The TENSOR is fully controlled by the OPUS software.

The HYPERION 1000 microscope is an accessory that may be connected to almost any type of FT-IR Bruker spectrophotometer. For fully non-destructive measurements, the TENSOR 27 spectrophotometer is connected to the HYPERION 1000 microscope. Solid specimens are usually processed in reflection.

Modern Technologies of Thin Films Deposition   Materials Research Forum LLC
Materials Research Foundations **39** (2018)   doi: http://dx.doi.org/10.21741/9781945291913

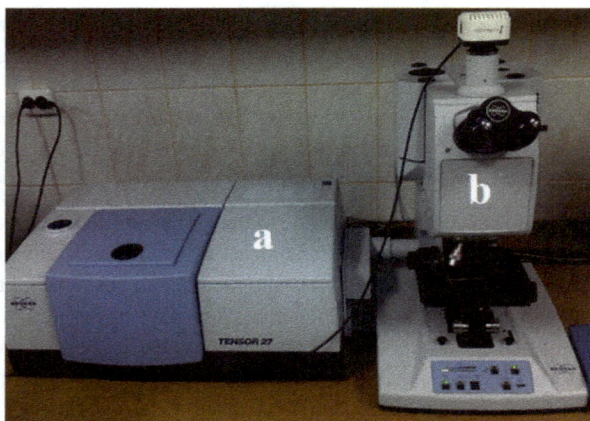

Fig. 4.11. *microFTIR Bruker Tensor 27 spectrophotometer:*
a - IR spectrometer, b - optical microscope.

The software is of the OPUS/VIDEO type for interactive video data acquisition. The processing may be done both in transmission and in reflection. The detector is of the MCT type cooled with liquid nitrogen (-196 $^{\circ}$C).

The spectral range is 600-7500 cm^{-1} and the measured area is optimized at a 250 μm diameter, but it may reach a minimum of 20 μm. The microscope is equipped with a 15X lens.

The disadvantage of the use of the system connected to a microscope is that the spectrum cannot be recorded within the 400 – 600 cm^{-1} range.

The advantages of the technique consist of the fact that there is a very rich database of spectra available and it allows the assessment of peak displacement towards certain wave numbers, such as the variation of their intensity. The technique also has other two advantages as compared to classical ones, which use dispersion in potassium bromide pill, namely: the avoidance of two errors, one brought about by uneven pill dispersion and another caused by the presence of hydroscopic water in the bromide, and the avoidance of fine powder sampling in the thin coating.

For testing purposes, the specimens are fastened on the microscope table by using a clamp and by focusing the light beam on the analyzed area. For comparative studies, several areas are differently analyzed or differently treated.

In experimental practice, specialists either use IR spectra [*Bentley, 1968; Nakamoto, 1997; Coates, 2000*], or rely on tables and diagrams with wavelengths specific to group vibrations [*Coates, 2000; Salzer, 2009*]. Since this technique allows access to the database based on wave numbers, the type of compound in the analyzed structure is automatically determined.

4.9   Tribological Tests

The friction force is the force occurring further to the interaction between objects when the two objects touch. Its direction is always opposite to the direction of movement and it has a dissipative nature [*Zhang, 2006, 2008; Iliuc, 1980; Czichos, 1978; Mateus, 2005*].

1st law of friction: The sliding friction force does not depend on the contact surface between two objects.

2nd law of friction: The friction force is directly proportional to the normal line on the surface and it depends on the nature of the surfaces touching each other by a material constant called friction coefficient, which is marked μ (miu). The friction coefficient is a physical quantity that does not have a unit if measurement, meaning it is adimensional.

The tribological tests were conducted using a ball-on-disk tribometer. The ball, 6 mm in diameter, is made of 100Cr6 steel, with mirror finished surface (Ra: 0.02 μm) and 62 HRC hardness.

The friction forces were measured using a variable linear differential transformer and recorded dynamically in a computer. The configuration and measuring components of the tribometer are shown in figure 4.12 [*Zhang, 2006, 2008; Iliuc, 1980; Czichos, 1978; Mateus, 2005*].

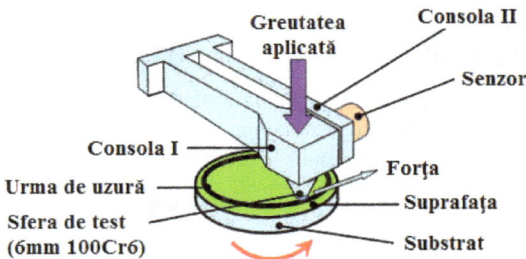

Fig. 4.12. *Basic configuration of a measuring part of the tribometer [Zhang, 2008].*

Figure 4.13 shows the device [*CSM Tribometers*].

Fig. 4.13. *Picture of the tribometer used.*

4.10 Determination of the Drawing Force

The action of phosphate films on the cold working process of steels has not been fully elucidated yet. However, it is known that the plastic deformation of the deposited crystals, which are very hard and at the same time fragile, does not occur by the sliding mechanism. Most crystal tips are thought to be broken during the plastic deformation process and turned into a very fine-grained powder. Together with the lubricant (applied before the deformation or not) on the adherent phosphate film, this powder forms a compact layer which allows sliding [*Calister, 1985; Geru, 1985; Baciu, 1996*]. Therefore, the powder formed maintains at the same time the connection with the metallic surface, playing a lubricant role, and also separates the contact surfaces.

In order to analyze the behavior during operation of phosphate films we chose small specimens, which were sampled by embossing. The phosphate-coated specimens underwent two successive drawing operations in order to create the part.

The behavior in operation of the phosphate films was analyzed by determining the value of the force required for the drawing of blanks treated by phosphate coating, as compared to the drawing of untreated blanks, in the case of the first drawing operation.

Although there is a multitude of analytical and experimental relations determined so far to calculate the drawing forces, neither of them takes or could take into consideration all the influence factors, hence the values of the forces obtained by the ratios are for orientation purposes. Therefore, most researchers prefer the experiment.

In order to determine the drawing force, we used a dynamic data acquisition system, which includes the fully equipped master unit, force transducers and displacement transducer.

The master unit is a Traveller 1 System, MUT-1 model, 1016-S type (Fig. 4.14-b), with the following main elements:

- 8 SG-2 tensometric amplification channels with 1kHz pass band;
- 4 optoisolated digital input channels, with external clock and trigger functions;
- 8 high signal direct analogical input channels;
- power supply from 12(24)Vcc battery or from embedded 220Vca/12(24)Vcc source;
- max. 100,000 Hz sampling frequency;
- Model EST-1 Basic software package for Windows 98/2000/XP;
- connection to computer by USB interface;
- full set of paired connectors and interconnection cables.

The force transducer with compression loan switch works based on resistive electrical strain-measuring technique (Fig. 4.14-c), with the following main characteristics:

- nominal load - 1000 kN;
- linearity deviation 0.1% C.S.;
- hysteresis deviation 0.1% C.S.;
- working temperatures: -20...+80°C;
- temperature-compensated range: 0...+60°C.

The displacement transducer, HS-100 model (Fig. 4.14-d), also works based on resistive electrical strain-measuring technique, with the following characteristics:

- measurement range: 102mm;
- resolution: infinite;
- linearity deviation: 0.2% C.S.;
- frequency response: 10Hz;
- working temperatures: -10...+60°C.

The equipment allows both the calibration of the transducers used and the acquisition of the data provided by them by means of the E.S.A.M. (Electronic Signal Acquisition Module) software, 3.0 version, with a frequency of up to 100 000 acquisitions per second.

Experimental results were obtained by means of the adequate equipment concerning the value of the drawing force depending on the punch stroke. The specimens were drawn both treated and untreated on the PHC 20 hydraulic press (Fig. 4.14-a) on which the drawing matrix I was placed.

Fig. 4.14. *PHC 20 hydraulic press (a) and Traveller 1 System acquisition equipment, MUT-1 model, 1016-S type (b); load cell, FN3000 1000 EH model (c); Displacement transducer, HS-100 model (d), computer with acquisition software (e) and power supply source (f).*

The equipment allows the acquisition of data provided by force and displacement transducers by means of the E.S.A.M. (Electronic Signal Acquisition Module) software, 3.0 version, the frequency used being 1 000 acquisitions per second.

The calibration of the load cell was done on a universal hydraulic press for 100 tf mechanical tests, manufactured in China. The calibration curve of the load cell (Fig. 4.15) represents a linear dependence between the force applied to the load cell and the output stress provided by the tensometric amplifier.

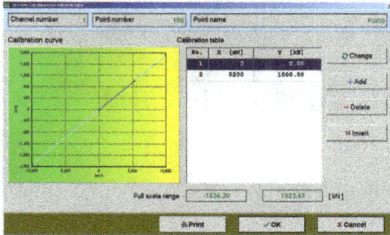

Fig. 4.15. *Calibration curve of the load cell.*

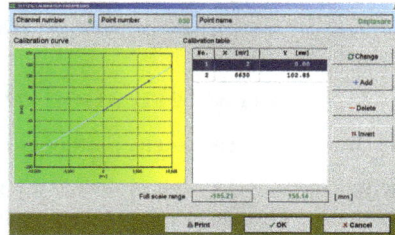

Fig. 4.16. *Calibration curve of the displacement transducer.*

Displacement transducer calibration was done by means of block gauges. The calibration curve of the displacement transducer (Fig. 4.16) is a linear dependence between the punch displacement and the output stress of the electronic tensometer.

Chapter 5

Characterization of Thin Phosphate Films

5.1 Indexing of Specimens Used in our Research

The specimens prepared according to the solutions shown in paragraph 2.4.1. are indexed in table 5.1, notations which will be used hereunder.

Table 5.1. Indexing of specimens depending on the composition of the phosphating baths.

Specimen	Composition of the bath
P0 – control specimen	Non-phosphate-coated sheet
P1	Standard phosphate-coating solution (SS) (80°C, 30 min) (see the composition in Table 3.5)
A	Standard phosphate-coating solution (SS) (90°C, 30 min)
B	Standard phosphate-coating solution (SS) (80°C, 15 min)
X1	SS + CoCl$_2$ (30 min) (see the composition in Table 3.6)
X2	SS + Ni(NO$_3$)$_2$ (30 min) (see the composition in Table 3.6)
X3	SS + CrO$_3$ (30 min) (see the composition in Table 3.6)
X4	SS + Bi(NO$_3$)$_3$ (30 min) (see the composition in Table 3.6)
Y1	SS + Ni(HCOO)$_2$ (30 min) (see the composition in Table 3.6)
Y2	SS + MgS$_2$O$_3$ (30 min) (see the composition in Table 3.6)
Y3	SS + Mn(NO$_3$)$_2$ (30 min) (see the composition in Table 3.6)
Y4	SS + Mg(NO$_3$)$_2$ (30 min) (see the composition in Table 3.6)
2A	SS (15 min) + (SS + Hexamethylenetetramine, 15 min) (see the composition in Table 3.6)
2B	SS + Hexamethylenetetramine (30 min) (see the composition in Table 3.6)
3A	SS (15 min) + (SS + Thiourea, 15 min) (see the composition in Table 3.6)
3B	SS + Thiourea (30 min) (see the composition in Table 3.6)
4A	SS (15 min) + (SS + Oak tannin, 15 min) (see the composition in Table 3.6)
4B	SS + Oak tannin (30 min) (see the composition in Table 3.6)

Modern Technologies of Thin Films Deposition   Materials Research Forum LLC
Materials Research Foundations **39** (2018)   doi: http://dx.doi.org/10.21741/9781945291913

In order to determine the best precipitation solutions, the standard solution was tested at different temperatures and for different reaction times. Among these specimens, three were selected as being representative, namely: P1 (standard phosphating solution at 80°C, 30 min reaction time), A (standard phosphating solution at 90°C, 30 min reaction time) and B (standard phosphating solution at 80°C, 15 min reaction time). Among these three specimens, specimen P1 was used as reference. The immersion time and precipitation temperature used for P1 were also used for the other specimens.

A much higher number of phosphating solutions were considered for the experiments, by adding to the standard solution insertion cations and additives with role of surfactants. For their creation, we considered the three biggest groups of reactivity, i.e. acid-base, redox and complexing, with their competitive balances. Among these, we selected the compatible systems for each block of cations and anions in the periodic system. Thus, Mg^{2+} was chosen for block "s", Bi^{3+} for block "p", Fe^{2+}, Cr, Mn, Co, Ni for block "d". Among the basic anions, the orthophosphate and nitrogen ions, among the compensation anions: chloride, thiosulfate and formate, and among the additives with role of surfactants: hexamethylenetetramine, thiourea and tannin.

Since, during the experiments, we noticed uniform and compact deposits in oxidant acid environments, in addition to the NO_3^- anion, we studied the role of chromic anhydride ($Cr^{VI}O_3$), which should be avoided in principal, but, considering that it is a very strong oxidant in the acid environment, we wanted to point out its role, as it forms very thin transparent films. Thus, we studied the influence of certain reductant anions, such as the formate anion, $HCOO^-$ ($HC^{II}O_2^-$) and thiosulfate, $S_2O_3^{2-}$ (S^{-I}-$S^{+V}O_3^{2-}$), the results of which are not what we expected.

A series of modern techniques were involved in the study of the characteristics of the deposited films and of their physical-mechanical behavior, which were grouped on the following tests:
-    morphology – OM, SEM, 3D optical profilometry;
-    composition – EDX, XRD and microFTIR
-    behavior – corrosion test, tribological and deformation tests.

5.2   Optical Microscopy (OM) Studies

The specimens were analyzed using the Zeiss Imager a1M optical microscope in the Scientific Research Laboratory of "Al. I. Cuza" University and the Leica DMRM microscope, used during the traineeship at the Belfort-Montbeliard Technology University in France. Thus, microphotographs of both phosphate-coated surfaces and of cross sections of thin films were taken.

Materials Research Forum LLC
doi: http://dx.doi.org/10.21741/9781945291913

5.2.1 Analysis of Phosphate-Coated Surfaces

Figures 5.1 – 5.5 show the microphotographs of the phosphate-coated surfaces after 12 months of preservation in the normal laboratory environment. The photographs show two different magnifications, namely 100 and 200X, and they were taken using two filters: BF – bright field and DF – dark field, which allow the taking of microphotographs with special contrast and real-life colors.

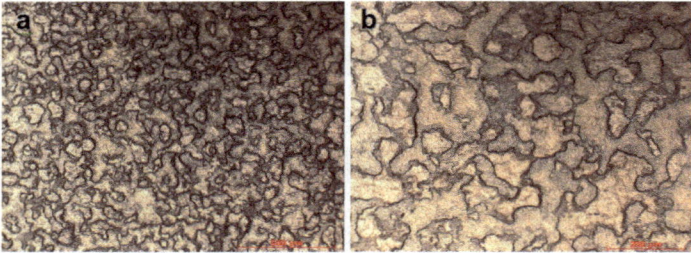

Fig. 5.1. *Image of the steel specimen (non-phosphtate-coated):*
a – 100X BF, b – 200X BF.

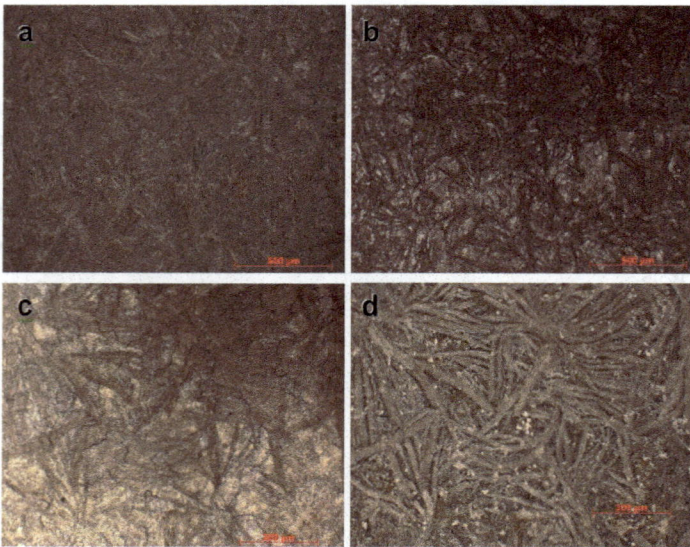

Fig. 5.2. *Micrography of the P1 phosphate-coated specimen:*
a – 100X BF, b – 100X DF, c – 200X BF, d – 200X DF.

Modern Technologies of Thin Films Deposition
Materials Research Foundations **39** (2018)

Materials Research Forum LLC
doi: http://dx.doi.org/10.21741/9781945291913

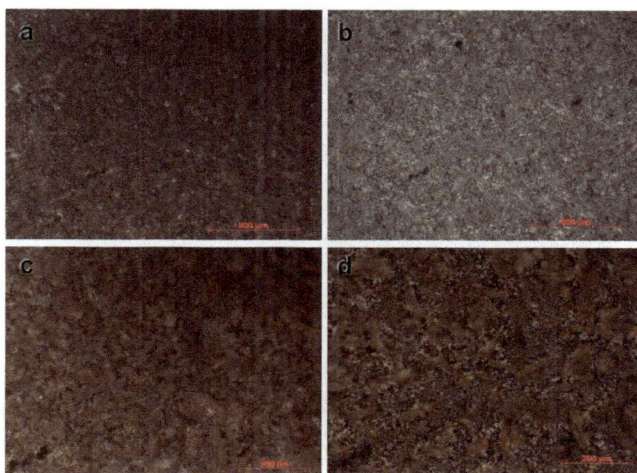

Fig. 5.3. *Microphotographs of phosphate-coated surfaces of specimen X2:*
a – 100X BF, b – 100X DF, c – 200X BF, d – 200X DF.

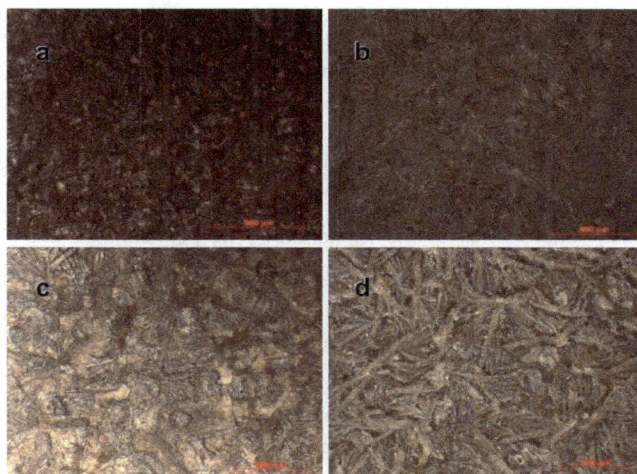

Fig. 5.4. *Microphotographs of phosphate-coated surfaces of specimen Y4:*
a – 100X BF, b – 100X DF, c – 200X BF, d – 200X DF.

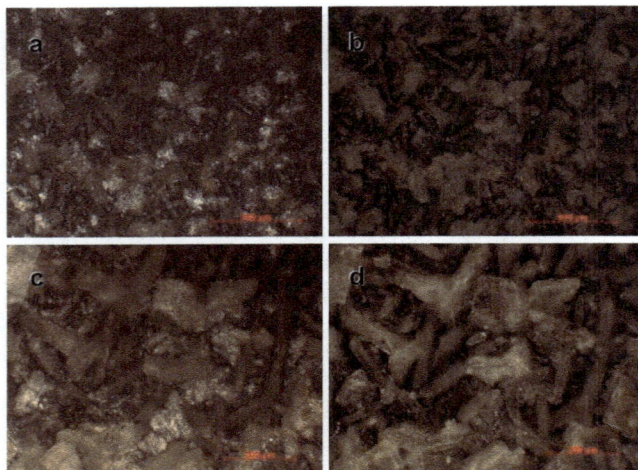

Fig. 5.5. *Microphotographs of phosphate-coated surfaces of specimen 2A:*
a – 100X BF, b – 100X DF, c – 200X BF, d – 200X DF.

We were able to follow the evolution of the surface structures during the twelve months they were kept in the laboratory, both against a dark field and against the bright field. Please note that oxidic spots occur and grow from the nucleation centers and in the pore and fissure areas [*Creus, 2000; Banczek, 2009*]; in less resistance films, the oxidic structures spread as oxidic spots or islands diffusing among dendritic structures.

When compared to the control specimen (non-phosphate-coated sheet) shown in figure 5.1, most tested solutions led to the formation of well-crystallized tree-like dendritic structures, with well represented active germination foci.

Please note that the dark field reveals clearly the details of the structures, whereas the shades represent insignificant chromatic deviations from the bright field microphotographs and from the real surface.

The films obtained in the standard solution are characterized by fine well-defined needle-shaped crystals, the morphology of which varies depending on the precipitation time and temperature. Thus, the crystallites in specimen P1 (30 min precipitation time at 80 °C) shows have the best-defined morphology –thin needle-shaped tree-like with few germination foci, A (30 min precipitation time at 90 °C) shows a finer structure with many germination foci, and B (15 min precipitation time at 80 °C) shows a similar structure to P1, yet it is highly porous, which supported the occurrence of corrosion stains.

As shown above, phosphate-coating in the presence of chromic anhydride resulted in the creation of very thin transparent films, the images of which are very close to those of the control specimen. Also, the coating on specimen Y2 is thin and its texture is close to that of the control specimen.

Among the films that are less stable in time, which led to the formation of diffuse oxidic stains and islands, we should distinguish specimens Y3 and 3B. In specimen one, the oxidation process is attributable to the susceptibility of cation Mn(II) to oxidation in the presence of air to Mn(III) and Mn(IV), respectively, which are more stable forms. Pitting formations also occur in specimens X1, which are attributable to Co(II) susceptibility to easy oxidation to Co(III), the stable form.

Difficultly soluble phosphates, obtained by conversion from zinc orthophosphate are generally stable to oxidation, even when they contain cations in inferior oxidation state, susceptible to oxidation to superior stable states. This explains why the films that contain Fe(II) ortho- and pyrophosphates are rather resistant to oxidation, due to their stabilization as congruent phosphophyllite ($Zn_2Fe(PO_4)_2 \cdot 4H_2O$). Moreover, in addition to achieving compact and even coatings due to the compatibility of orthophosphates with the formed pyrophosphates, the presence of the Ni(II) cation, which is stable to oxidation, confers great stability to oxidation. Among the two Ni(II)-based solutions, specimen X2 (Fig. 5.3) has a very fine and compact structure, whereas Y1 forms coarse dendritic structures due to the strongly reductant formate anion.

Crystalline structures were obtained in Mg^{2+} solutions in the presence of the thiosulfate anion, yet they result in porous structures. Also, structures with well-defined crystals, in which the two types of crystallites are revealed, namely needle-shaped fine - resulting from the primary phosphating process in standard solution, followed by insertion of polyhedral phosphate crystallites formed during the secondary process in the presence of hexamethylenetetramine (Fig. 5.5).

Initially, the addition of hexamethylenetetramine, tannin and thiourea did not have the expected results, as the crystallites were very small, grown together, with pores and fissures. Thus, we first performed a light phosphate-coating in the standard solution for 15 minutes, followed by a second 15 minute-phosphating process in the presence of additives. Please note that the results obtained for tannin and thiourea were similar to those obtained for hexamethylenetetramine, yet the dendritic structures were smaller, uneven and not compact.

5.2.2 Cross-Section Analysis of Thin Films

Figures 5.6, 5.7 show the cross sections of the thin films obtained with the help of the two microscopes immediately after thin film deposition (a – 50X, BF and b – 100X, BF) and after about 12 months of preservation in the normal laboratory environment (c 100X, DF), only for a series of representative specimens.

The Zeiss Imager a1M microscope allowed us to measure the thickness of the films against the cross section, against a dark field, which reveals the shades of the structures with the smallest chromatic deviation.

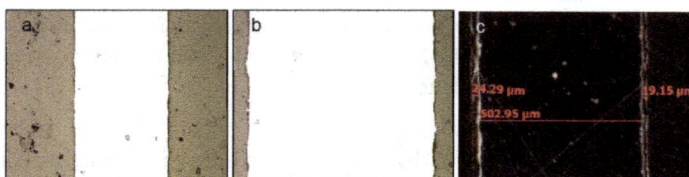

Fig. 5.6. *Micrography in cross-section for specimens P1:*
a – 50X BF, b – 100X BF, c – 100X DF.

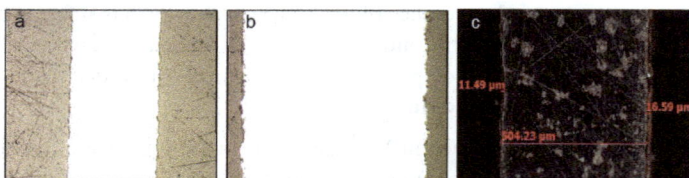

Fig. 5.7. *Cross-section microphotographs of specimen X2:*
a – 50X BF, b – 100X BF, c – 100X DF.

Specimens P1, A and B obtained in standard solutions, but at different temperatures and precipitation time, had coating thickness well correlated with the precipitation time and temperature (P1 19...24μm; A – 12...47μm; B – 14...16μm).

Note that the solutions that led to thin compact and even films resistant to corrosion (analyzed by macro and microscopy) are correlated with the measurements in the cross-section microphotographs.

Thus, specimen X2 – with Ni(II) (Fig. 5.7) has a 11.5 to 16.5µm thick film, specimen 2A – hexamethylenetetramine has a 11.5 to 19 thick film, and specimen 4A – with tannin between has a 13 to 16 thick film.

The films deposited in magnesium solutions are uneven and porous, ranging from 7 to 20µm, whereas those with cobalt are even but high porous, ranging from 24 to 28µm.

5.3 SEM-EDX Analysis

Scanning electron microscopy together with X-ray spectroscopy and the two SE and BSE detectors represent one of the most important material science and engineering techniques, as it allows the recording of images of the surface of the specimens, with 100 to 100,000X magnification, of the electron dispersion spectrum, based on which elemental composition is assessed in gravimetric or atomic percentages, and of atom distribution of the analyzed surface.

Except for the reference specimens A and B (using P1 as the only reference), for the other 14 specimens we recorded 100X, 500X and 1000X images, atom distribution on the analyzed surface at 500X and EDX spectrum, based on which we assessed the elemental composition shown in table 5.2.

Thus, figure 5.8 shows SEM microphotographs of the control specimen (sheet prior to phosphate coating), the composition of which is described in paragraph 3.3.

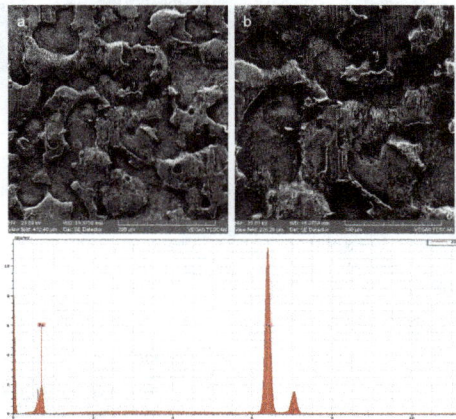

Fig. 5.8. *SEM microphotographs (SE - secondary electrons) of steel prior to phosphate coating: a - 500X SE, b – 1000X SE and c – EDX spectrum.*

Both the 500X and 1000X images reveal the surface structure which is the result of the rolling processes. Since the other components of the DC01 steel are below the deviation of the device, the EDX spectrum only reveals the presence of iron.

Figures 5.9-5.12 show the 100X, 500X and 1000X surface images, together with element distribution on the surface at 500X and the EDX spectrum for the reference specimen P1, X2, Y4 and 2A, respectively.

The standard steel part phosphating solution allows the creation of thin films made of thin tree-like needle-shaped dendritic crystallites, with visible germination or nucleation foci. Atom distribution on the surface clearly renders the two processes of structural reformation through sequential coprecipitation, the first subtractive process forming iron phosphate (reddish) and the second additive process forming zinc phosphate (greenish).

Since for the creation of thin films the same precipitation conditions are applied as for specimen P1, the same main characteristics will be analyzed in the specimens: shape of crystallites, germination foci distribution, distribution of basic cations precipitating as phosphate, then of insertion cations and the other elements resulting from additions or from the compensation anions, which will be presented comparatively.

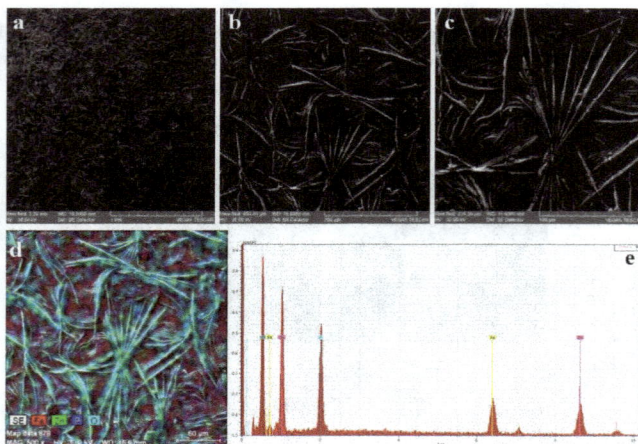

Fig. 5.9. *SEM (SE) images and EDX spectrum data of specimen P1:*
a – 100X, b – 500X, c – 1000X, d – atom distribution on the surface; e – EDX spectrum.

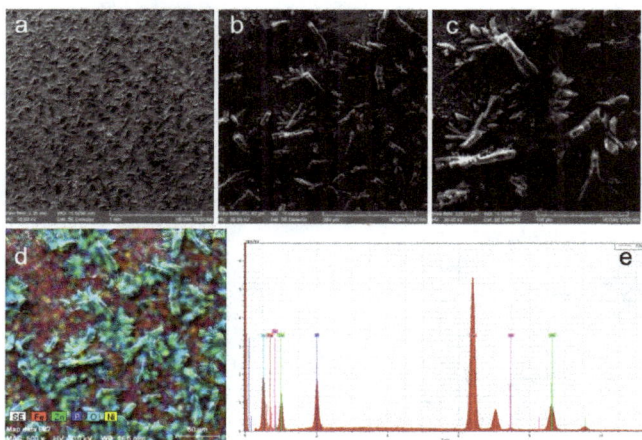

Fig. 5.10. *SEM (SE) images and EDX spectrum data of specimen X2:*
a – 100X, b – 500X, c – 1000X, d – atom mapping; e – EDX spectrum.

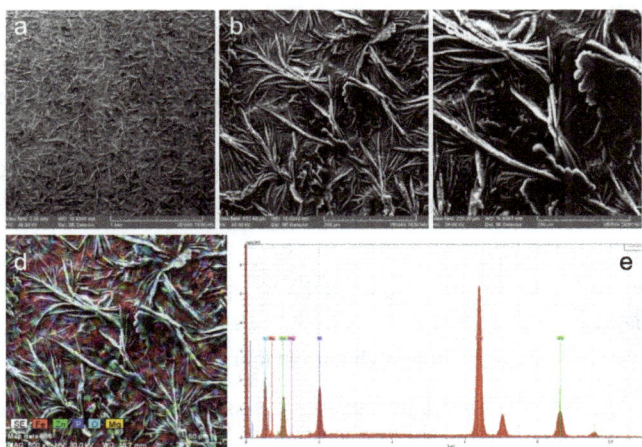

Fig. 5.11. *SEM (SE) images and EDX spectrum data of specimen Y4:*
a – 100X, b – 500X, c – 1000X, d – atom distribution on the surface; e –EDX spectrum.

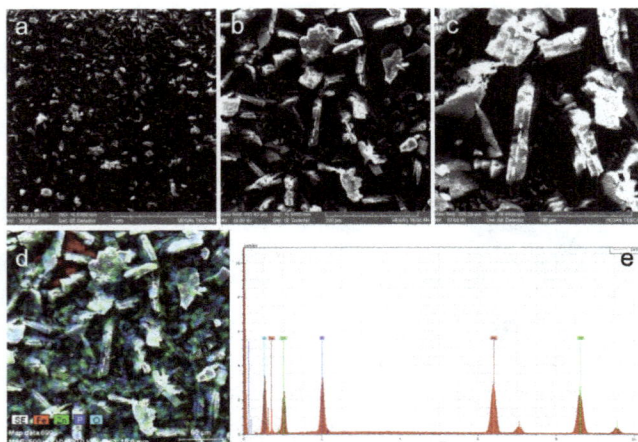

Fig. 5.12. *SEM (SE) images and EDX spectrum data of specimen 2A:*
a – 100X, b – 500X, c – 1000X, d – atom distribution on the surface; e –EDX spectrum.

Table 5.2. shows the elemental composition in atomic percentages assessed in the EDX spectra of the samples, which allow the detection in the surface structures of basic phosphate cations and of cations inserted by addition, plus a series of atoms from the addition components with role of surfactants or compensation anions.

The visual analysis of the SEM microphotographs and of the atom distribution on the surface allows the grouping of the 15 sets of specimens in the following series:

- needle-shaped dendritic crystallites: P1, X1, Y4, 2B, 3A and 4A;
- coarse compact crystallites: X2, 2A and Y1;
- attenuated or missing crystallites: X3, X4, Y2 and Y3;
- scale- or shell-like crystallites with uneven morphology: 3B and 4B.

Five of these specimens are the most interesting, three from the first series (P1, X1 and Y4) and two from the second series (X2 and 2A), in which the Zn(II) phosphate coating achieved during the secondary process has uneven distribution, with small gaps, in which the Fe(II) phosphate coating created during the primary process is noticeable.

Table 5.2. *Element composition determined using the EDX spectra.*

Element Specimen	Composition (% at)											
	Fe	Zn	P	O	Ni	Co	Mg	Bi	Cr	Mn	N	S
P1	7.86	12.94	12.13	67.07	-	-	-	-	-	-	-	-
X1	22.84	10.91	13.15	51.45	-	1.65	-	-	-	-	-	-
X2	22.05	9.40	12.77	55.31	0.47	-	-	-	-	-	-	-
X3	74.62	0.32	0.22	24.56	-	-	-	-	0.27	-	-	-
X4	58.36	3.31	6.50	30.85	-	-	-	0.98	-	-	-	-
Y1	8.12	13.41	15.11	62.23	1.13	-	-	-	-	-	-	-
Y2	68.43	2.95	3.20	16.06	-	-	3.03	-	-	-	-	6.33
Y3	44.19	2.65	10.65	39.30	-	-	-	-	-	3.20	-	-
Y4	41.45	6.64	8.87	40.02	-	-	0.61	-	-	-	2.41	-
2A	10.41	14.58	15.30	59.71	-	-	-	-	-	-	-	-
2B	32.51	11.27	13.47	42.75	-	-	-	-	-	-	-	-
3A	35.47	8.97	12.08	43.48	-	-	-	-	-	-	-	-
3B	39.77	13.74	9.04	37.44	-	-	-	-	-	-	-	-
4A	35.50	8.68	11.67	44.05	-	-	-	-	-	-	-	-
4B	21.54	14.79	15.34	48.32	-	-	-	-	-	-	-	-

In specimens P1, X1, Y4 and 2B, one may notice congruent phosphophyllite structures $(Zn_2Fe(PO_4)_2 \cdot 4H_2O)$, with even distribution; by contrast, in specimens X2 and 2A they are inserted between the Fe(II) and Zn(II) phosphate; moreover, in specimen X2 the

Ni(II) phosphate is clearly revealed next to the phosphophyllite, with which it is structurally compatible; in specimen 2A, the formation of hexamethylenetetramine with Fe(II) and Zn(II) supported the formation of polygonal crystallites distributed over the phosphophyllite.

When analyzed in general, the morphology of surface crystallites of thin Zn(II) and Fe(II) orthophosphate films obtained by coprecipitation in aqueous orthophosphoric and nitric acid-based solutions, which involve competitive substitution and addition processes [*Sandu, 2011, 2012g and 2012h*], creates an uneven coarse structure with long branchy or flat crystallites.

In the presence of the surfactant (hexamethylenetetramine - HT, tannin and thiourea) these crystallites become shorter, thicker and more compact, having the shape of flattened structures (scales, flakes, shells, etc.).

On specimen immersion in the three solutions with added surfactants, the deposited coatings had different crystallite morphology. When hexamethylenetetramine was added, non-adherent polyhedral crystals of the coordinative complexes formed with HT and water as ligand also occurred in addition to thin dendritic crystallites (Fig. 5.12). The tannin solution supports the formation of fine uneven lenticular crystallites, whereas the thiourea solution leads to the formation of bigger shell-shaped crystallites.

Since the three surfactants support component solubility in the phosphating solution, the added solutions do not easily generate crystallization germs in the iron substrate. For these reasons, the specimens were initially immersed in the standard phosphating solution for 15 minutes, then they were immersed for another 15 minutes in solutions with additives, i.e. coprecipitation components and surfactants.

After their immersion for 15 minutes in the standard solution and another 15 minutes in the solution with additives, the one with hexamethylenetetramine allowed the creation of a coarse compact polyhedral superimposed crystallite film (Fig. 5.12), whereas the others led to the creation of fine dendritic structures.

Specimens obtained in solutions with addition of magnesium cations, in the presence of two different compensation anions, thiosulfate and nitrate, the former generating very thin almost transparent structures, and the latter generating prolonged dendritic structures.

The precipitation solution in which chromic anhydride was added had an interesting behavior. It is known to be a very strong oxidant, which may lead metal cations in superior stable oxidation states (Fe(III), Mn(III,IV), Co(III), Ni(III) etc.) the phosphates of which are not structurally compatible with Fe(II) and Zn(II) orthophosphates and have a solubility product higher than them, thus forming very thin transparent films.

Materials Research Forum LLC
doi: http://dx.doi.org/10.21741/9781945291913

According to the EDX data (table 5.2) the (Fe+Zn+other coprecipitation cations)/P molar ratio for the phosphate coatings, which vary in the sequence:

Y1 (1.5) < 2A (1.63) < P1 (1.71) < 4B (2.37) < X2 (2.50) < X1 (2.69) < 2B (3.25) < 3A (3.68) < 4A (3.79) < Y3 (4.7) < Y4 (5.5) < 3B (5.95) < X4 (9.64) < Y2 (23.25) < X3 (342)

is very well correlated with film compactness and thickness.

As one may note in specimens X2 and Y4, zinc is substituted by nickel and magnesium, respectively, and in specimen 2A the higher concentration of zinc to the detriment of iron is due to the action of the surfactant, which passivizes the iron substrate.

In all four cases the O/P ratio is approximately 4, which corresponds to the stoichiometry of orthophosphate, the excessive oxygen is attributable to the bonds with orthophosphates and to the water in the crystal-hydrates, and for those with a 3 to 4 ratio it is attributable to the bonds with pyro and metaphosphates.

These specimens will be analyzed hereunder in order to identify the chemical nature of the crystalline microstructures by XRD and microFTIR.

5.4 XRD Diffractometric Analysis

Figures 5.13 – 5.17 show the XRD spectra for specimens P1, X1, X2, Y4 and 2A. The SEM-EDX analyses revealed special structural characteristics of the latter from the morphological and compositional viewpoint. The spectra were obtained by using a BRUKER diffractometer, found in the Laboratory for Material, Process and Surface Studies and Research of the Belfort-Montbéliard Technology University of France.

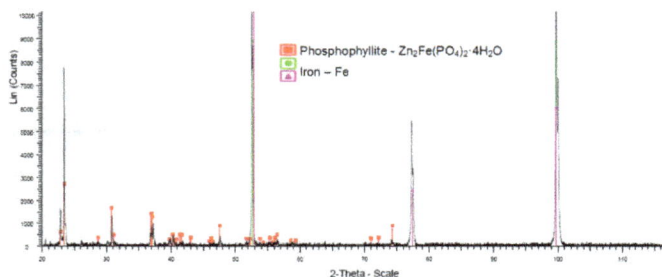

Fig. 5.13. XRD spectrum for specimen P1.

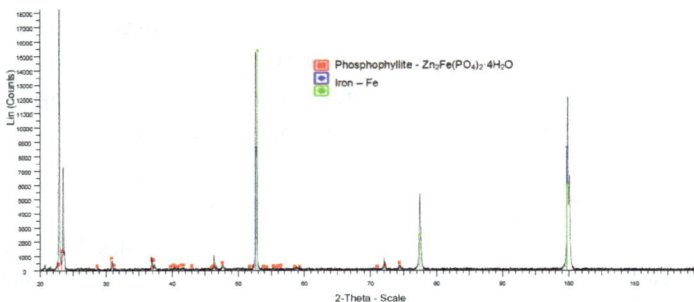

Fig. 5.14. *XRD spectrum for specimen X1.*

Based on these spectra we determined the chemical formula of the basic congruent and revealed the allotropic forms by the parameters of the elementary cell and of the crystallization system. All the specimens contain basic congruent structures scattered in the coating, in the form of phosphophyllite, $Zn_2Fe(PO_4)_2 \cdot 4H_2O$ (double Zn and Fe phosphate tetrahydrate) with monoclinic crystallization system and monoclinic primate elementary cell (P21/c) with the following parameters: a = 10.377Å, b = 5.086Å, c = 10.559Å, $\alpha = 90°$, $\beta = 121.1°$ and $\gamma = 90°$.

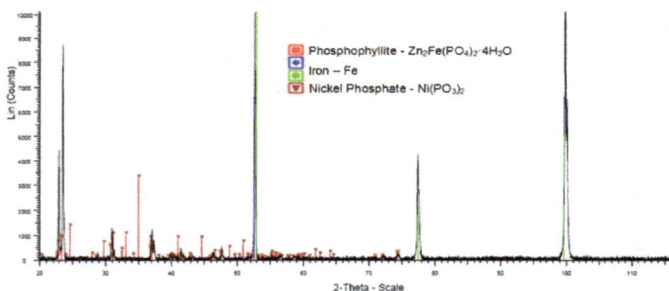

Fig. 5.15. *XRD spectrum for X2.*

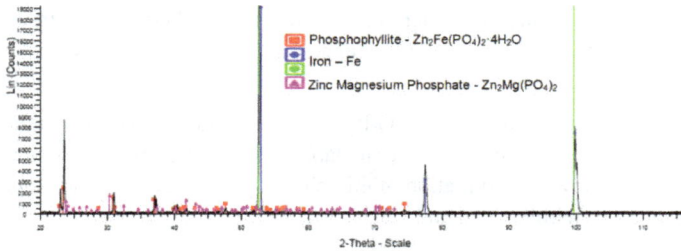

Fig. 5.16. *XRD spectrum for specimen Y4.*

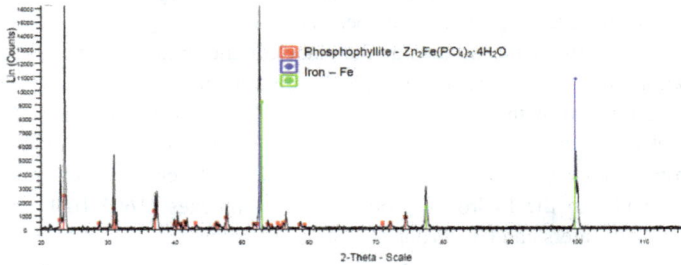

Fig. 5.17. *XRD spectrum for specimen 2A.*

In specimen X2, in addition to phosphophyllite, there is also nickel pyrophosphate $(Ni(PO_3)_2)$, which is structurally compatible with the first one, the last one being included in the monoclinic system, with central elementary cell, with the following parameters a = 11.086 Å, b = 8.227 Å, c = 9.832 Å, α = 90°, β = 112.74° and γ = 90°.

In specimen Y4, in addition to phosphophyllite, there is also double anhydrous zinc and magnesium phosphate $(Zn_2Mg(PO_4)_2)$, also compatible with phosphophyllite, being crystallized in monoclinic system, with primitive monoclinic elementary cell (P21/n), with the following parameters: a = 7.569 Å, b = 8.355 Å, c = 5.059 Å, α = 90°, β = 94.95° and γ = 90°.

The morphology and distribution of the crystals for the two specimens X2 and Y4 are totally different, and this difference is accounted for precisely by the big difference between the β angle of elementary cells, as the other parameters provide good structural compatibility in crystallite formation. In the former case, coarse even structures, distributed throughout the phosphophyllite- and nickel pyrophosphate-containing surface, are formed, whereas in the latter case, filiform needle-shaped structures, with phosphophyllite and double zinc and magnesium phosphate evenly distributed, are formed.

These data may also account for the O/P combination ratios over 4 in table 5.2. The presence of nickel in the form of metaphosphate, structurally compatible with orthophosphates, allows the formation of thin phosphate-type ceramic coatings, with very good resistance to environmental and mechanical factors.

5.5 MicroFTIR Spectroscopy Analysis

The compounds formed as a result of the phosphate coating processes applied to the analyzed specimens are supported by the peaks of the specific group vibrations, both the δ deformation ones (in plane and outside the plane), in the small wave number range, and the v vibration ones. The FTIR spectra reveal only structures in the form of compounds and not in the form of metallic structures. Thus, the FTIR spectra (Fig. 5.18-5.24) reveal the local peaks specific to the orthophosphate groups (920-950 cm^{-1} and 1054-1086 cm^{-1}), to the pyrophosphate groups (616-627 cm^{-1} and 973-1000 cm^{-1}), to the nitrate groups (1436-1467 cm^{-1}), to the hydroxo- group in the complexes (1605-1640 cm^{-1}), to the physically bound waters and HO$^-$ groups (3230-3565 cm^{-1}).

The technique used has a drawback, more precisely it limits registration in the 4000-600cm^{-1} range, and also a significant advantage, which consists of the fact that it is a non-destructive technique, which does not require specimen sampling and processing by the KBr pellet preparation technique, which in its turn would introduce two significant sources of deviations: specimen dispersion irregularity in KBr and KBr humidity.

As concerns the waters group, in the form of a wide band in the 3200 to 3600 cm^{-1} range, two groups are distinguished in the spectra of the analyzed specimens, namely the hydroxocomplex group (HO$^-$), 3520-5565cm^{-1} and the physically bound water group, 3230-3450cm^{-1}.

Most products resulting from the phosphate coating processes contain orthophosphate, pyro and metaphosphate structures, which are clearly revealed by individual peaks, except for specimens X3, X4, Y1, Y2 and Y3. X1 exhibits only the vibrations specific to pyro and a single vibration specific to ortho; the same vibration is also found as a single peak in specimens Y2 and Y3.

The absence or strong diminution of these peaks is accounted for by the absence of the phosphate coating or the forming of oxide structures (below 700 cm^{-1}) and a single type of very thin orthophosphate, transparent to IR radiation (about 950 cm^{-1}).

Specimens X3, X4, Y1 and Y2 reveal clearly only the surface waters and HO⁻ anion in hydroxocomplexes.

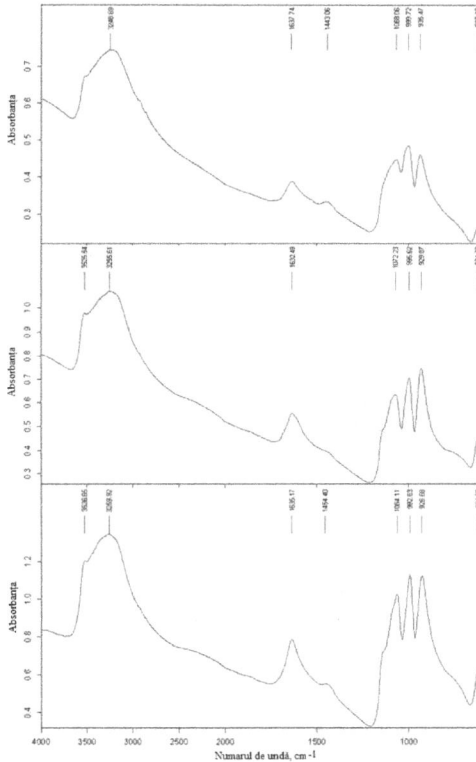

Fig. 5.18. *FTIR spectra for specimens P1, A and B.*

The spectra of specimens P1, A and B (Fig. 5.19), which were obtained in standard solutions, the only variables being the temperature and precipitation time, were very similar, and only revealed the main types of functional groups: phosphates (ortho, pyro), nitrates, hydroxides and physically bound waters by well-individualized peaks. Among these, P1 was the specimen with the best IR spectroscopy, optical microscopy and scanning electron microscopy findings, which made it the reference specimen.

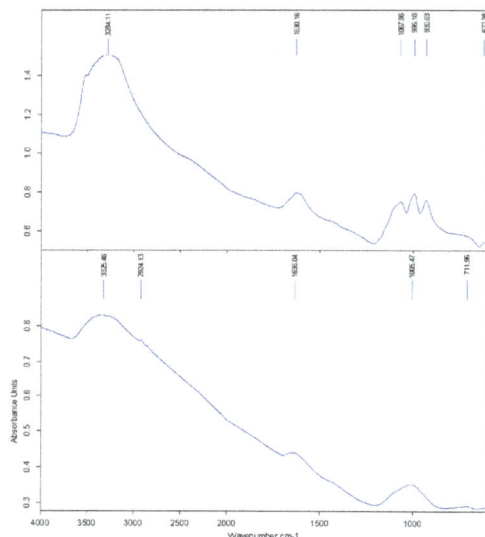

Fig. 5.19. *FTIR spectra for specimens X2 and Y1, with Ni(II) added.*

The specimens obtained by additivation with other cations and other surfactants will be classified according to the additivation system.

As concerns the specimens to which nickel was added (Fig. 5.20), we noted two coatings with completely different behaviors: X2 (in which $Ni(NO_3)_2$ was used – oxidant system) with all the peaks clearly revealed, except for the nitrate anion (very attenuated), close to the spectrum of the reference specimen P1 and Y1 (in which $Ni(HCOO)_2$ was used – a reductant system) with all the peaks strongly diminished. The formate anion, a very strong reductant agent, in the presence of the nitrite anion, did not allow the formation of difficultly soluble Fe(II), Zn(II) and Ni(II) phosphates.

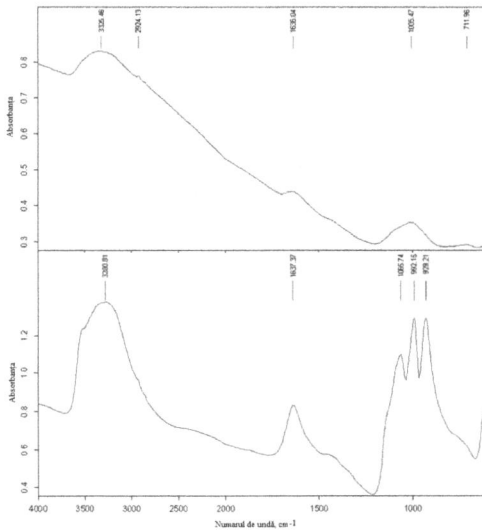

Fig. 5.20. *FTIR spectra for specimens Y2 and Y4, to which Mg(II) was added.*

The specimens to which magnesium was added (Fig. 5.20), by means of two anions with very different redox character, namely NO_3^- (strong oxidant) and S_2O_3 thiosulfate, (strong reductant agent), have spectra very close to Ni(II) spectra. Specimen Y4 (to which $Mg(NO_3)_2$ was added) has a spectrum similar to specimen X2, whereas specimen Y2 (to which MgS_2O_3 was added) has a spectrum similar to specimen Y1.

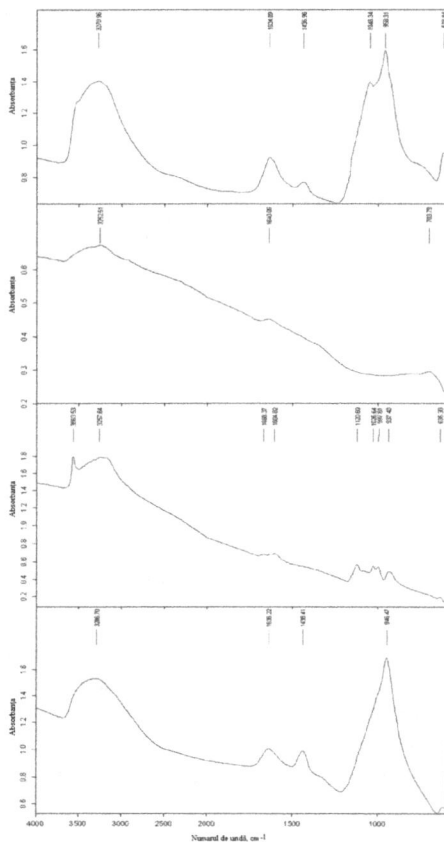

Fig. 5.21. *FTIR spectra for specimens to which cations are added:*
X1 – to which $CoCl_2$ was added, X3 – to which CrO_3 was added,
X4 – to which $Bi(NO_3)_2$ was added and Y3 – to which $(NO_3)_2$ was added.

Specimens X1 (to which $CoCl_2$ was added – Cl^- weak reductant agent) and Y3 (to which $Mn(NO_3)_2$ was added – oxidant system) also have similar spectra, the only difference being the fact that the peaks only occur for ortho, whereas the NO_3^- peak is best revealed in other specimens (Fig. 5.21).

Specimens X3 (CrO$_3$) and X4 (Bi(NO$_3$)$_2$), respectively, only have spectra with very attenuated peaks, the only peak occurring only in specimen X4. Chromic anhydride is a very strong oxidant, which destabilizes the minimum Fe(II) oxidation state and turns it into Fe(III), whose solubility product in relation to phosphate anion is much higher than that of Fe(II) phosphate and structurally incompatible to phosphophyllite. In specimen X4, Bi(III) is a cation with big ionic radius, as compared to Zn(II) and Fe(II) radii, with a tendency to stabilization as Bi(III) more in the form of difficultly soluble oxide-hydroxides and less in the form of phosphates. The local phosphate peaks belong to Fe(II) and Zn(II) attenuated by the presence of bismuth oxide-hydroxides as thin superposed coatings.

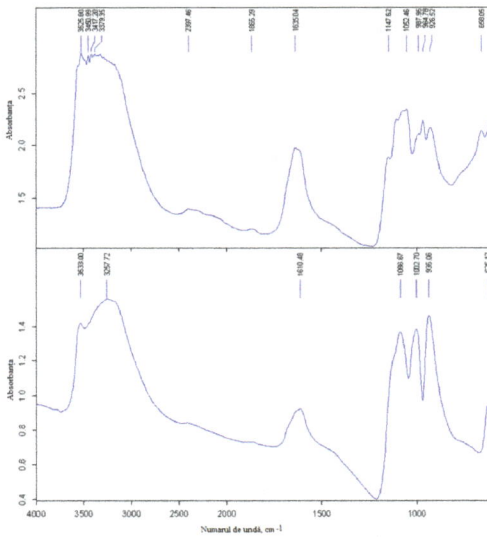

Fig. 5.22. *FTIR spectra for coatings in which hexamethylenetetramine was added: 2A and 2B.*

As concerns the addition of surfactants (hexamethylenetetramine, thiourea and tannin), the spectra for the two series of coatings deposited sequentially (15 minutes in standard solution followed by another 15 minutes in solution to which an additive was added and 30 minutes in solution to which an additive was added) are somewhat different. Thus, specimens 2A (two-stage sequential coating) and 2B (single-stage coating in solution with additive), to which hexamethylenetetramine was added, have a complex spectrum, which includes the specific groups of this additive: methyl (660 and 2393 cm^{-1}), C-N group (1869 cm^{-1}) and complexed NH_2 group (2850-2930 cm^{-1} and 3174 cm^{-1}), in addition to peaks specific to ortho, pyro and hydroxo groups, the peak of the nitrate ion being very attenuated. Poorly soluble phosphates are grown together with the hexamethylenetetramine complexes, by forming coarse overlapping structures.

In specimen 2B, obtained in a single stage, the structures belonging to transitional cation complexes in the standard solution (Zn(II) and Fe(II)) with hexamethylenetetramine are hardly visible, as opposed to phosphate structures, which are very visible.

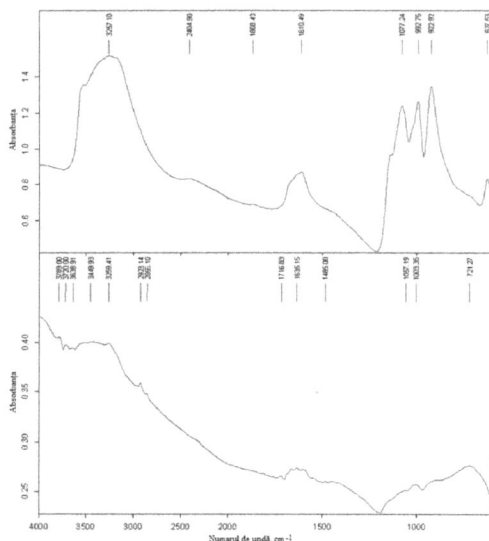

Fig. 5.23. *FTIR spectra for coatings to which thiourea was added: 3A and 3B.*

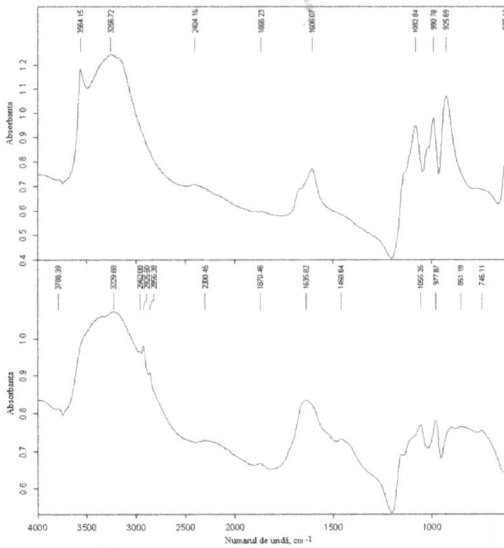

Fig. 5.24. *FTIR spectra for coatings to which tannin was added: 4A and 4B.*

When additive adding is a two-stage sequential process, specimens 3A and 3B to which thiourea was added (Fig. 5.23) and specimens 4A and 4B to which tannin was added (Fig. 5.24) have almost identical spectra. However, the spectra obtained after additives were added to the solution do not reveal the presence of phosphates and hydroxocomplexes, but only the presence of transitional cation complexes with tannin and thiourea, respectively.

According to the IR spectroscopy data on the intensity of the peaks specific to phosphate groups, the analyzes specimens, in the forms of ortho and pyrophosphate, in addition to congruent phosphophyllite and other poorly soluble products, crystallized interstitially in the dendrites or in other coarse compact and irregular structures, may be grouped in the following series in relation to the coating of the reference specimen P1: close spectra (X2, Y4, 2A, 2B, 3A, 4A) with well individualized peaks; spectra with less marked peaks (X1, X4, Y3, 4B) and the last series, spectra with very poor and strongly attenuated peaks (X3, Y1, Y2 and 3B).

Modern Technologies of Thin Films Deposition   Materials Research Forum LLC
Materials Research Foundations **39** (2018)   doi: http://dx.doi.org/10.21741/9781945291913

5.6 Study of Resistance to Corrosion in NaCl 0.1M Solution

The five most representative specimens, chosen among the sampled specimens by SEM-EDX and XRD analysis, underwent a corrosion test and they were then analyzed comparatively by scanning electron microcopy and microFTIR. In order to allow better understanding of the nature of the coating deposited, table 5.3. shows the compositions of the phosphating baths for the five specimens.

Table 5.3. *Composition of phosphating baths.*

Bath composition	P1	X1	X2	Y4	2A
H_3PO_4 solution 98% (mL/L)	8.16	8.16	8.16	8.16	8.16
Zn (g/L)	4.00	4.00	4.00	4.00	4.00
HNO_3 solution 60% (mL/L)	2.60	2.60	2.60	2.60	2.60
NaOH (g/L)	0.75	0.75	0.75	0.75	0.75
$NaNO_2$ (g/L)	0.45	0.45	0.45	0.45	0.45
$Na_3P_3O_{10}$ (g/L)	0.05	0.05	0.05	0.05	0.05
$CoCl_2 \cdot 6H_2O$ (g/L)	-	12.00	-	-	-
$Ni(NO_3)_2 \cdot 6H_2O$ (g/L)	-	-	12.00	-	-
$Mg(NO_3)_2 \cdot 6H_2O$ (g/L)	-	-	-	15.00	-
Hexamethylenetetramine (g/L)	-	-	-	-	2.00

We conducted corrosion tests both on the DC01AM steel blank, which was considered the standard specimen, and on the five specimens chosen, with optimum phosphate films. All the experiments were conducted at room temperature (25 °C), by using an electrochemical cell with 3 electrodes connected to an Autolab PG STAT 302N potentiostat (Metrohm Autolab) with Nova 1.6 software. The reference electrode, against which the potential values were expressed, was made of saturated calomel (ECS), a flat platinum electrode was used as auxiliary electrode and the working electrodes were the specimens themselves.

The cyclic voltammograms (CV) of the specimens were recorded within the -1.2÷1.8V range at a v=100 mV/s scanning rate, and they revealed point corrosion, as hysteresis specific to this type of corrosion occurred on the back scanning of the potential [*Nemtoi, 2007 and 2010*].

In the standard specimen in 0.1M NaCl, the passive range (Fig. 5.25) is much narrower (-0.55÷-0.05V) and the sudden anodic current increase occurs similarly as in specimen 2A, yet the current reaches higher values (≈8mA), then its flattening occurs (Fig. 5.26). On the back scanning of the potential, the current decreases to lower potential values than the sudden leap during forward scanning; the hysteresis also occurs in this case, yet it is narrower than in specimen 2A.

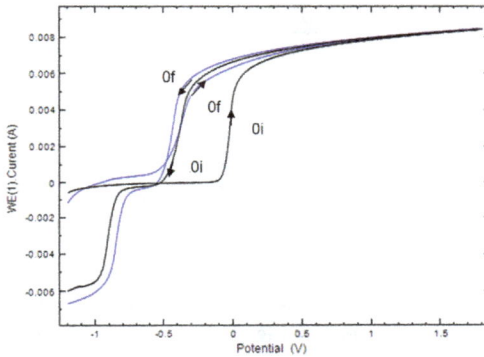

Fig. 5.25. *CV for the standard specimen immersed in 0.1MNaCl solution (St$_i$, forward and back scanning) and after 5 hours (St$_f$).*

The cyclic voltammogram for specimen 2A within the -1.2÷0.8V range shows that the specimen is virtually passivized within this range, not being attacked by 0.1M NaCl; by contrast, the current suddenly peaks at 0.8V up to 6mA, as mentioned in literature [*Sanchez, 2006 and 2007*], and then it becomes flat, probably due to the coating deposited on the surface of the specimen. The hysteresis that we referred to above occurs on back scanning, the repassivation field does not virtually reoccur, and the passing from anodic to cathodic current occurs at -0.445 V. In time, the passive field is reduced, the proof being that when specimen 2A is kept in 0.1M NaCl solution its resistance to corrosion decreases considerably.

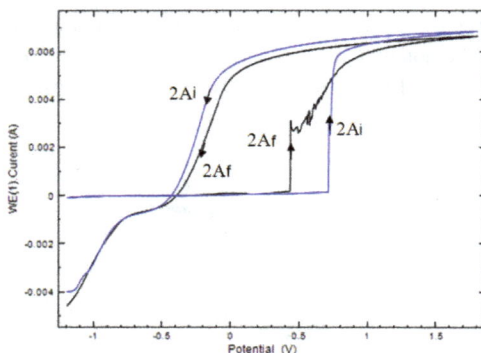

Fig. 5.26. *CV for 2A (2A$_i$ forward and back scanning) after 12 hours (2A$_f$).*

The cyclic voltammograms of specimens 2A and X2 are shown in figure 5.27 where the passive field for X2 is virtually absent. Figure 5.28 shows the cyclic voltammograms of specimen X1, which does not exhibit any passive field and for which the back scanning does not initially reveal any point corrosion, and for specimen Y4, which one may say that it exhibits a very narrow passive field.

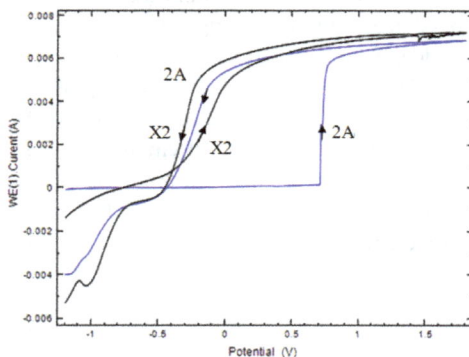

Fig. 5.27. *CV for 2A and X2 in NaCl 0.1M solution.*

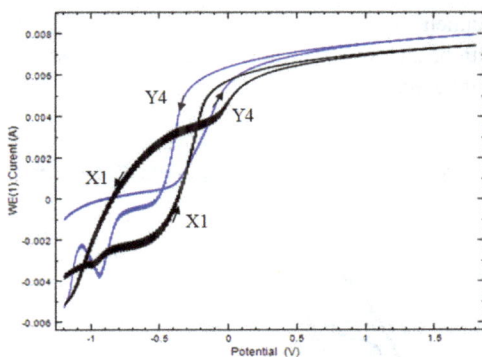

Fig. 5.28. *CV for X1 and Y4.*

Specimen P1 has an interesting behavior within the -0.834÷-0.300V range, after which the current suddenly peaks and then become flat, unlike specimen P1, which was previously degreased and pickled (P1d) and which exhibits no passivation field but only the flattening (even a slight decrease) of current within the -0.558÷-0.363V potential range, as shown in figure 5.29.

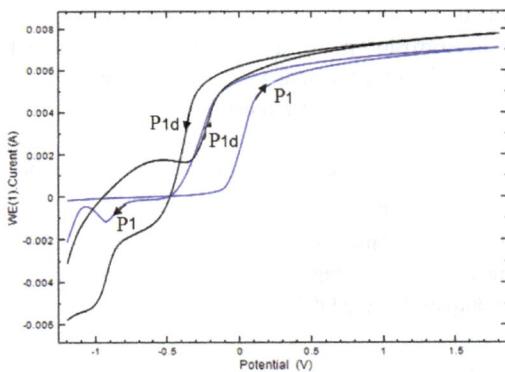

Fig. 5.29. *CV for P1.*

As compared to specimen P1, we may say that Y4 is more stable and less influenced by the 0.1M NaCl solution, although the passivity field of P1 is bigger, as shown by the voltamograms in figure 5.30.

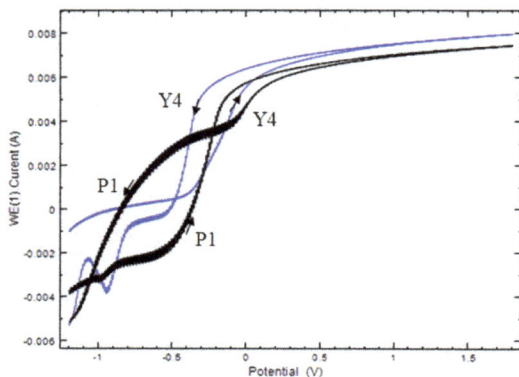

Fig. 5.30. *CV for P1 and Y4.*

In order to determine the corrosion parameters, we drew the linear voltamograms within the ±150 mV range by considering the open circuit potential (OCP) of each specimen, determined before the drawing of the linear voltamogram. The OCP values are shown in Table 5.4, whereas the corrosion potential calculated by the Nova 1.6, $E_{cor,calc}$ software takes into consideration the changes that may occur during LV drawing, yet it is not considerably different from OCP; the difference is however slightly bigger for specimens Y4 and P1, as shown in Table 5.4. When calculating the corrosion rate (expressed by the penetration speed), iron oxidation was thought to occur at $Fe(OH)_2$, as supported by literature data [*Ningshen, 2009*].

Figure 5.31 shows Evans diagrams for the 6 specimens analyzed in aqueous 0.1 M NaCl solution, by drawing the linear voltamograms at 1 mV/s scanning rate over a 300 mV range (±150 mV as compared to the OCP of each specimen).

Materials Research Forum LLC
doi: http://dx.doi.org/10.21741/9781945291913

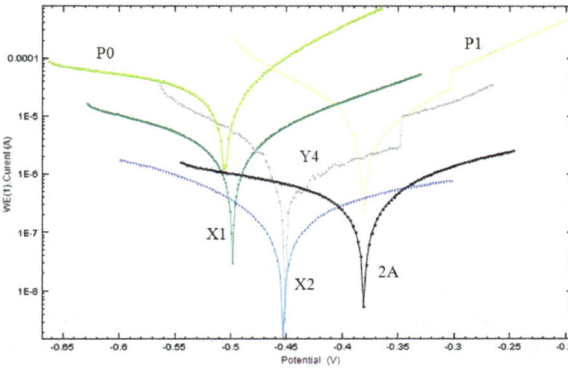

Fig. 5.31. *Evans diagram for the six samples in a NaCl 0,1 M solution.*

Table 5.4. *Corrosion parameters of specimens analyzed in aqueous NaCl solutions.*

specimen	-OCP (mV)	b_a (mV/dec)	b_c (mV/dec)	$E_{cor,calc}$ (mV)	j_{cor} ($\mu A/cm^2$)	v_{cor} (mm/an)	R_p (Ω)
NaCl 0.1M							
X2	450	148.96	240.22	-451	0.0564	$0.6554 \cdot 10^{-3}$	187890
2A	395	292.21	172.48	-388	0.1384	$1.6078 \cdot 10^{-3}$	108420
Y4	414	96.35	166.30	-458	0.6542	$7.6022 \cdot 10^{-3}$	12897
X1	506	170.48	140.30	-507	0.9475	$11.01 \cdot 10^{-3}$	11235
P1	348	90.72	116.58	-390	2.9159	$33.882 \cdot 10^{-3}$	2016.7
P0	513	351.94	103.09	-511	9.3373	$108.50 \cdot 10^{-3}$	1181.1

The corrosion parameters were determined by using the Tafel lines in Evans diagrams, shown in Table 5.4.

According to the date in table 5.4, the nickel specimen has the highest resistance to corrosion among the specimens with protective coating, whereas P1 is the most vulnerable. Also, according to CV for specimen Y4, we found that within a relatively short time from immersion (12 h) the resistance to corrosion in this corrosive environment decreases significantly (from 108420 Ω to 3272.9 Ω), and the resistance to corrosion of specimen X1 drops even lower than that of the standard specimen after degreasing and pickling, probably due to the fact that these operations also remove the protective coating. For the standard specimen, we also determined the corrosion parameters in a more corrosive environment (0.25 M NaCl) where, as one may notice in table 5.4, the resistance to corrosion is the lowest.

Due to its high surface wetting and dispersion ability, hexamethylenetetramine supports the formation of a thick regular and compact coating with fully grown dendrites, which improves resistance to corrosion.

Thus, the coprecipitation process first results in the formation of a fine ceramic coating of iron phosphate and zinc ($Zn_2Fe(PO_4)_2 \cdot 4H_2O$ - phosphophyllite), in which nickel pyrophosphate $Ni(PO_3)_2$ is then inserted. As compared to other coprecipitation processes in the aqueous system, which only use phosphate anions in acid environment, by means of oxidant accelerators like nitrate and other surface additives (with wetting and dispersion ability), the procedures are susceptible to competitive processes based on acid-base balances in the presence of certain cations in stable oxidation states (for instance, Mg^{2+}, Ca^{2+}, etc.) or of certain polyvalent cations with low stability in their inferior oxidation state, which, in the presence of air oxygen, oxidants and light, reach higher oxidation states (Mn^{2+}, Cr^{2+}, Fe^{2+}, Ni^{2+} etc.).

Theoretically, for instance, cation Fe^{2+} (just like Mn^{2+}, Cr^{2+} and Ni^{2+}), which is a weak instable base, easily turns to Fe^{3+} in the presence of air oxygen. Indeed, in a weak acid environment, Fe^{2+} is easily oxidized and turned into la Fe^{3+}, yet its good compatibility with cation Zn^{2+} in the presence of phosphate anion, with which it forms a poorly soluble ceramic phosphophyllite film, with high chemical inertia in redox and acid-base balances, is achieved with a very difficultly soluble film-forming component, which adheres to the substrate and is uniformly deposited. Moreover, in the presence of cation Ni^{2+}, which forms with the pyrophosphate ion a very stable ceramic compatible with the insertable structure of phosphophyllite, it increases coating stability, as it becomes more compact and has no thermal or rheological cracks.

5.6.1 *Comparative Study Before and After SEM Corrosion*

Figure 5.32. shows a comparison between the SEM views before and after corrosion.

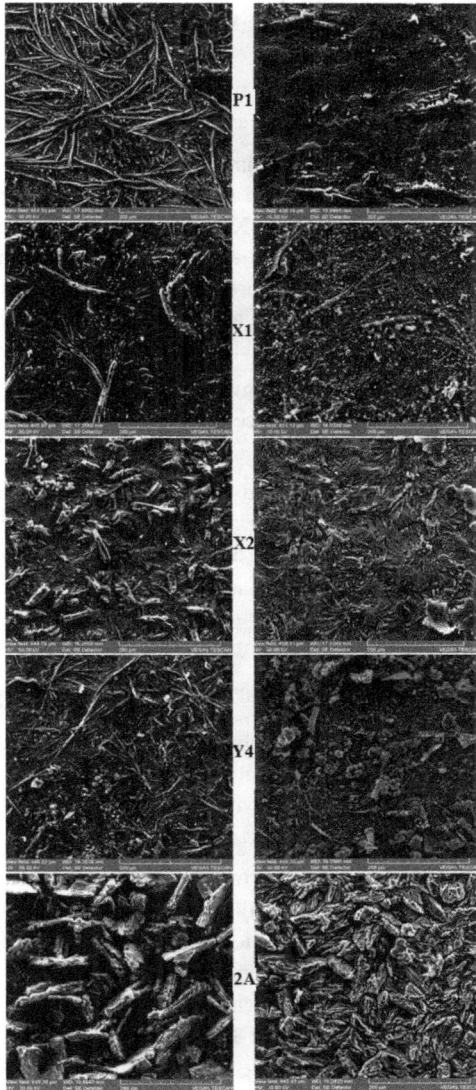

Fig. 5.32. SEM views of specimens before and after corrosion.

Figure 5.32 shows the morphology and evenness differences between the surface structures of the five specimens.

Thus, unlike specimen P1, which has thin and elongated tree-like dendrites with evenly scattered crystallization foci, specimens X1 and Y4 have scarce and irregular dendrites.

After corrosion, the structures of specimens X2 and 2A, with totally different morphologies, are compact, even and with no cracks, and the steel substrate does not show, like in the other specimens (P1, X1 and Y4). Well-developed 3D dendritic structures thicken the coating by vertical increase (Sandu, 2012a and 2012b). Since the structures of specimens X2 and 2A are more compact and the crystallites enjoy a more uniform distribution, the pores and gaps are absent. Specimens P1, X1 and Y4 have areas where the steel support is visible, which makes them susceptible to corrosion attack.

The pores and cracks allow the corrosive environment to come in direct contact with the steel support, which means that resistance to corrosion is very poor (Creus, 2000; Banczek, 2009).

The SEM views of specimens P1, X2 and Y4 after corrosion reveal the fact that the protective coating is destroyed to a large extent, whereas the coating on specimens X2 and 2A is compact, has no pores or cracks, and the steel support is not visible. The poor resistance to corrosion of specimens P1, X2 and Y4 may be accounted for by the thin and elongated dendrites, which have very big surfaces in contact with the corrosive environment.

5.6.2 *Comparative Study Before and After Corrosion by microFTIR*

The compounds formed after the phosphate-coating processes applied to the five specimens and the traces left by the corrosion test are also confirmed by the FTIR spectra (Fig. 5.33-5.37), which are well correlated with the EDX and XRD data.

Thus, the FTIR spectra reveal the peaks specific to the orthophosphate group (919-940 cm^{-1} and 1054-1085 cm^{-1}), to the pyrophosphate group (616-624 cm^{-1} and 973-1000 cm^{-1}), to the nitrate group (1438-1467 cm^{-1}), to the hydroxo- group in the complexes (1605-1636 cm^{-1}), to the physically bound waters and to the HO⁻ groups (3249-3560 cm^{-1}). The spectrum of specimen 2A (with hexamethylenetetramine) includes the specific methyl groups (660 and 2393 cm^{-1}), for the C-N group (1869 cm^{-1}) and the complexed NH_2 group (2850-2930 cm^{-1} and 3174cm^{-1}).

Figures 5.33 – 5.37 show by comparison the microFTIR spectra of the five specimens, before and after corrosion.

Materials Research Forum LLC
doi: http://dx.doi.org/10.21741/9781945291913

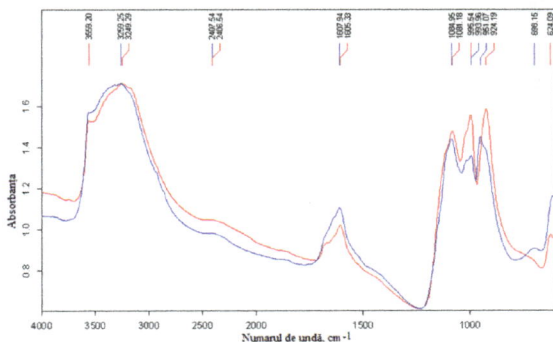

Fig. 5.33. *microFTIR spectrum of specimen P1:*
red – before corrosion and blue – after corrosion.

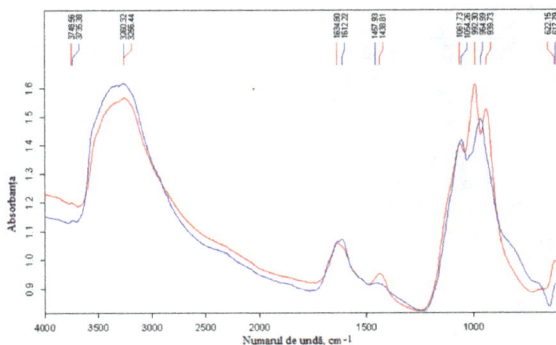

Fig. 5.34. *microFTIR spectrum of specimen X1:*
red – before corrosion and blue – after corrosion.

Materials Research Forum LLC

doi: http://dx.doi.org/10.21741/9781945291913

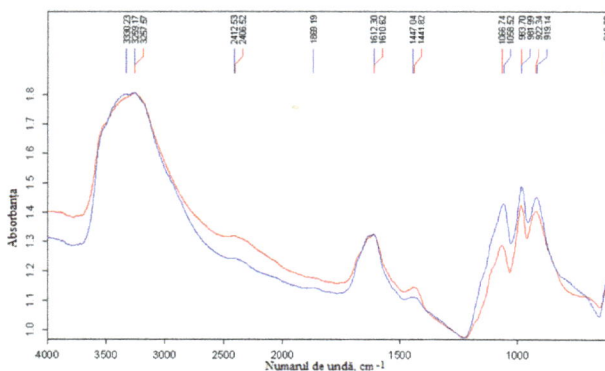

Fig. 5.35. *microFTIR spectrum of specimen X2: red – before corrosion and blue – after corrosion.*

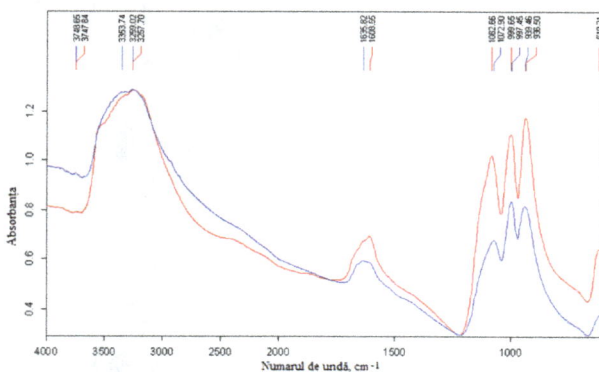

Fig. 5.36. *microFTIR spectrum of specimen Y4: red – before corrosion and blue – after corrosion.*

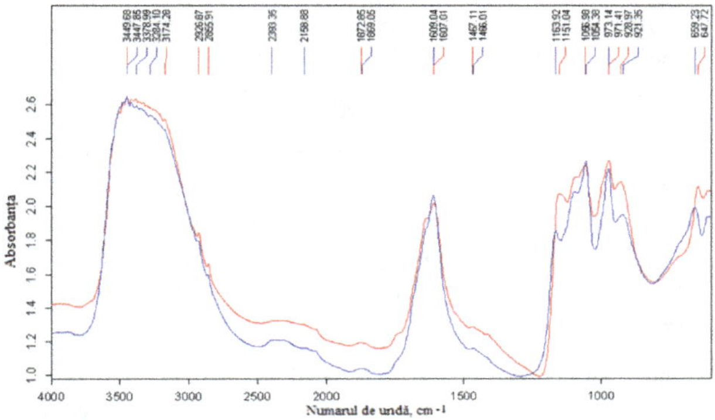

Fig. 5.37. *microFTIR spectrum of specimen 2A:*
red – before corrosion and blue – after corrosion.

After the corrosion tests, the specimens had similar FTIR spectrum, with specific groups, which were attenuated (I/I_0) or displaced ($\Delta\lambda$) as a result of the corrosion processes (table 5.5).

Table 5.5. *Comparative data concerning changes in the specific group peaks of the five specimens before and after corrosion.*

Specimen	Group								Specific range
	Orthophosphates		Pyrophosphates		NO_3^-		HO^-/H_2O		
	$\Delta\lambda$	I/I_0	$\Delta\lambda$	I/I_0	$\Delta\lambda$	I/I_0	$\Delta\lambda$	I/I_0	
P1	26.88	0.92	0.00	1.25	17.75	1.05	2.61	*1.09	* 1607.94-1605.33cm^{-1}
	3.77	0.98	-1.58	0.87			9.96	**1.00	** 3259.25-3249.29cm^{-1}
X1	25.26	0.98	4.36	0.92	19.12	0.95	-22.58	*1.02	* 1634.80-1612.22cm^{-1}
	-7.47	1.01	20.29	0.84			3.88	**1.04	** 3260.32-3258.44cm^{-1}
X2	-3.20	1.03	0.00	1.02	6.22	0.97	1.68	*1.00	* 1612.30-1610.62 cm^{-1}
	-8.22	1.10	-1.71	1.05			1.60	**1.00	** 3259.17-3257.57cm^{-1}
Y4	-2.96	0.64	0.00	0.73	7.31	0.92	27.27	*0.86	* 1635.82-1608.55cm^{-1}
	9.76	0.67	-2.20	0.74			1.28	**1.00	** 3259.02-3257.70cm^{-1}
2A	-7.62	0.90	11.51	0.94	-1.10	0.91	2.03	*1.02	* 1609.04-1607.01cm^{-1}
	-2.60	1.01	1.73	0.98			20.13	**0.98	** 3284.10-3263.97cm^{-1}

$\Delta\lambda$ - local peak displacement (λ corroded - λ non-corroded)

I/I_0 – peak intensity attenuation (I/I_0 = I corroded / I non-corroded)

The data on the resistance to corrosion of the five specimens reveal the following decreasing sequence of the specimens: X2 > 2A > Y4 > X1 > P1.

The data in table 5.5 confirm that specimen X2 enjoys better protection to corrosion than the other specimens, as in most specific group peak changes the small exceptions are due to the stability of the hydroxo- and tetraamino- complexes for the metallic cations involved in the phosphating processes. The best correlation was noticed in the I/I_0 ratios for both peaks of the orthophosphate group and for the nitrate peak, and the $\Delta\lambda$ displacement for the second group of orthophosphate-specific peaks and for the first pyrophosphate group.

Thus, the attenuation and displacement effects allow the determination of resistance to corrosion, being correlated both with the SEM-EDX data on compactness and composition, and with the corrosion rate based on the tests performed.

The attenuation and displacement effects in the microFTIR spectra allow the determination of the resistance to corrosion, being correlated both with the SEM-EDX data on compactness and composition, and with the corrosion rate based on the tests performed.

5.7 3D Profilometric Analysis

In other applications than resistance to corrosion, the roughness of the phosphate coatings is a very important characteristic. It is possible to obtain thin rough coatings resistant to corrosion due to the compactness and evenness in the interaction area with the substrate, as well as thicker rough coatings, yet less resistant to corrosion due to the existence of pores and cracks.

In general, rough compact films with high crystallites play an important role in plastic cold and hot working lubrication. The lubrication capacity may be the result of these crystallites, which, when they break, they form an additional lubricant, mobile and very thin film, or it may be due to the carrying cavity system which allows the inclusion of film-forming lubricants based on graphite, molybdenite or other colloidal dispersion. Hence the importance of this study.

Each specimen underwent 3D profilometric analysis performed by means of an optical AltiSurf 500 profilometer, which is a multi-sensor optical system (Fig. 4.4). This allows the drawing of several graphical representations, namely:

- the topography of the surface of the films in two graphical systems (with structure height variation grid based on colors and by stereomicrography based on shadows);

- roughness assessment in the topography variation graph per vector;

- particle distribution per unit of surface;

- morphological particle distribution on the analyzed surface with description of their number, average height and surface.

They are shown in figures 5.38-5.41 which correspond to the analyzed specimens.

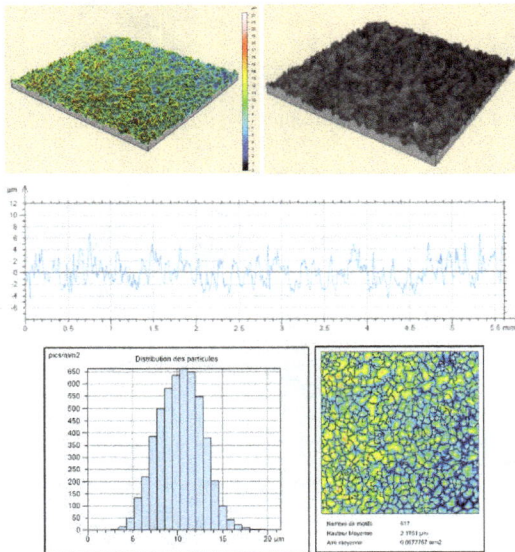

Fig. 5.38. 3D profilometry data for P1.

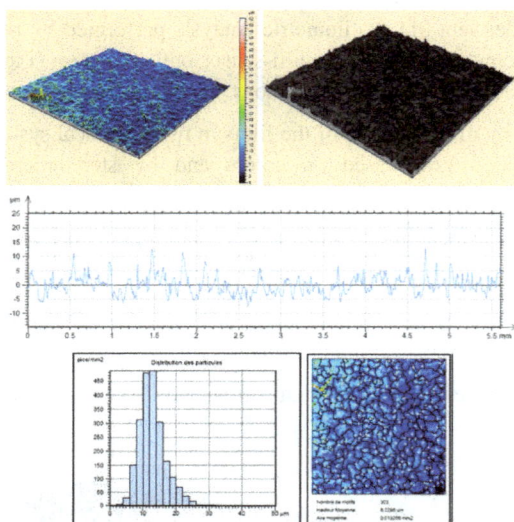

Fig. 5.39. *3D profilometry data for specimen X2.*

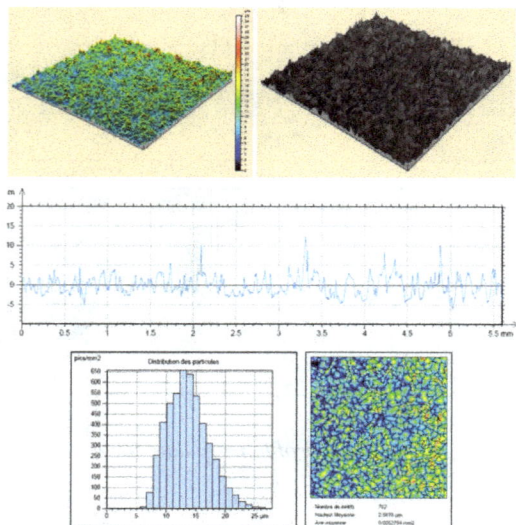

Fig. 5.40. *3D profilometry data for specimen Y4.*

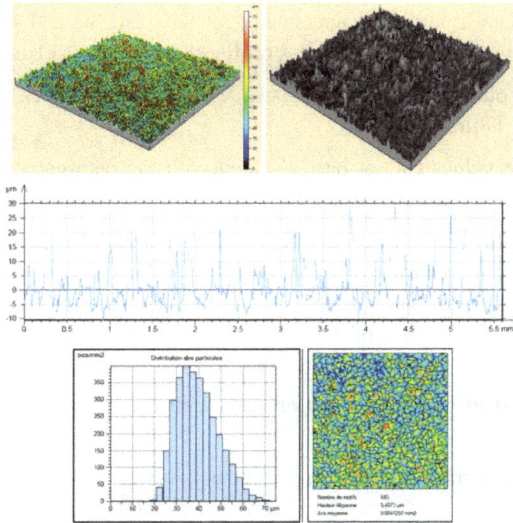

Fig. 5.41. *3D profilometry data for specimen 2A.*

The microtopographic representations using the two systems clearly reveal the evenness and compactness of the coatings. Here are the most even specimens of all the specimens under survey: P1, X2, X3, X4, Y2, 2B, 3A and 4A, whereas the rest of the specimens have uneven granulometries, in which one may distinguish a series of higher structures, which may play a special role in the cold plastic working processes.

Whereas the first two representations are meaningful only as an iconographic system rendering surface topography, which allows qualitative assessment, the other three representations allow quantitative assessments.

Among these, roughness, which was determined as an average by multiple tests (n = 8) on a 5.6 mm length, by using optical 3D profilometry, is represented as numerical data, more precisely as arithmetic mean deviation of the roughness profile. The data are shown in table 5.6.

Table 5.6. *Specimen roughness.*

Specimen	P1	X1	X2	X3	X4	Y1	Y2	Y4	2A	2B	3A	3B	4A	4B
Ra (μm)	1.49	1.88	2.21	1.83	1.21	4.33	1.63	1.65	5.65	1.44	1.67	2.45	1.76	2.83

According to the data in table 5.6, specimens X2, Y1, 2A, 3B and 4B have the highest values of the arithmetic mean of roughness profile (higher than 2.11μm).

The values of the other specimens (P1, X1, X3, X4, Y2, Y4, 2B, 3A and 4A) range between 1.21 and 1.88μm.

At first sight, these values cannot be correlated with the data gathered by microscopic analysis in cross section, which reveals thin film evenness and porosity.

Thus, in the first high roughness group (X2, Y1, 2A, 3B and 4B), specimens X2 and 2A, which are at the end of the range, have the best film compactness and evenness. In the second group (P1, X1, X3, X4, Y2, Y4, 2B, 3A and 4A), the specimen deposited in the standard solution, the roughness of which tends towards the lowest end of the range, has the best compactness and evenness.

The granulometric distribution analysis of surface structures was conducted on a 4mm^2 (2x2mm) area.

Table 5.7 shows the normal distribution parameters, namely the minimal diameter (D_{min}), the maximal diameter (D_{max}) and the mean diameter (D_{med}) of the particle, with the highest particle number distribution (No.pic/D_{med}).

As shown in table 5.7., specimens P1, X2, X3, Y2, Y4 and 2B have a narrow granulometric range, between 5 and 27μm, which corresponds to a very fine structure. On the other hand, the granulometric range of the other specimens (X1, X4, Y1, 2A, 3A, 3B, 4A and 4B) is very wide, between 7 and 70μm, and corresponds to a coarse structure.

Table 5.8 shows the morphological characteristics of surface structures, i.e. the number of particles on the 4mm^2 area, their mean area and height.

Table 5.7. *Granulometric distribution.*

Specimen	P1	X$_1$	X$_2$	X$_3$	X$_4$	Y$_1$	Y$_2$	Y$_4$	2A	2B	3A	3B	4A	4B
D$_{min}$ (μm)	4	8	4	5	4	15	5	7	20	5	7	12	10	10
D$_{max}$ (μm)	17	50	26	25	35	70	25	25	65	27	30	55	25	45
D$_{med}$ (μm)	11	15	14	16	21	33	11	13	35	14	22	28	19	22
No.pic/ D$_{med}$	650	430	480	265	130	440	420	650	400	210	170	250	730	560

Table 5.8. *Morphological characteristics of the surface structures*

Specimen	P1	X_1	X_2	X_3	X_4	Y_1	Y_2	Y_4	2A	2B	3A	3B	4A	4B
No. part/ 4mm^2	517	195	303	186	400	581	366	762	845	339	393	507	457	908
H_{med} (μm)	2.17	7.48	6.23	3.84	3.67	10.26	2.90	2.57	9.4	3.09	2.80	7.80	2.46	6.48
S_{med} x 10^{-3} (mm^2)	7.77	20.61	13.26	21.61	10.05	6.86	10.98	5.27	4.7	11.86	10.23	7.92	8.79	4.37

These characteristics are correlated with the granulometric distribution described above.

5.8. Tribological Tests

The tribological tests were performed by means of a CSM ball tribometer. The specimens were fastened on the holder which rotates at 20cm/s on a 5 mm radius and travels 100m (500 seconds, about 3200 revolutions), under a 2N load. The tribometer head is a 100C6 steel ball (Ra: 0.02 µm, HRC 62) fastened on a bracket, just like in figure 4.8.

The tribological tests were performed only on some of the specimens (P1, X1, X2, Y1, Y2, Y3, Y4, 2A, 2B, 3A, 3B, 4A and 4B), which had even structures, but, from the morphological point of view, they proved somewhat different as concerns vertical crystallite development; during plastic deformation, crystallites may play an important lubrication role as they form a thin mobile film made up of pieces broken from these crystallites.

Figures 5.42-5.44 show the friction coefficient variation of these specimens.

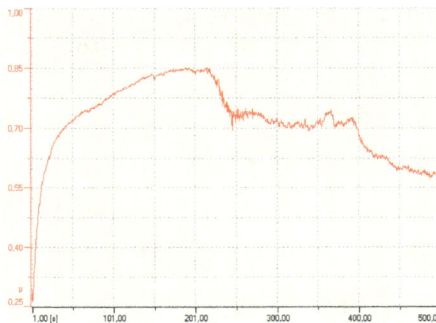

Fig. 5.42. *Friction coefficient variation for specimen P1.*

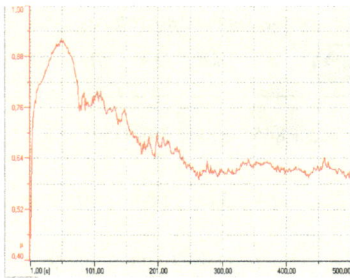

Fig. 5.43. Friction coefficient variation for specimen X2.

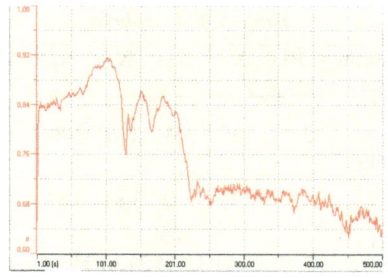

Fig. 5.44. Friction coefficient variation for specimen 2A.

Based on these diagrams and in order to assess the friction behavior of the phosphate films, we chose a series of qualitative data which allow comparing specimen behavior and its correlation with other characteristics.

Here are some these data:

- maximum specimen coefficient (μ_{max}),;
- number of peaks recorded before film destruction.
- time when the maximum coefficient occurs (t_{max});
- time when the coefficient suddenly drops (t_{min}) – which is correlated with the moment when the protection layer is destroyed to a large extent;
- final value of the friction coefficient of the mechanically deformed deposit (μ_f), at the end of the analysis, when there virtually is no more protection coating – this value is close to the value of the steel substrate (~0.6).

Table 5.9. Values of the data collected further to the tribological tests.

Specimen	Maximum coefficient, μ_{max}	Number of peaks	μ_{max} t_{max} (s) time	Time when the coefficient suddenly drops t_{min} (s)	Final value of the friction coefficient, μ_f
P1	0.85	1	190-210	230	0.58
X1	0.92	1	95	120	0.60
X2	0.92	3	50; 105; 150	200	0.61
Y1	0.92	3	30-45; 80; 120	150	0.55
Y2	0.72	1	120	170	0.55
Y3	0.87	1	25	90	0.55
Y4	0.93	3	60; 100; 170	230	0.58
2A	0.92	3	105; 150; 180	220	0.63
2B	0.87	3	40-50; 100; 150	180	0.58
3A	0.87	3	30; 125; 155	195	0.62
3B	0.86	3	25; 90; 120	160	0.52
4A	0.92	1	110	190	0.58
4B	0.84	2	35; 70	85	0.52

Some of the specimens have several friction coefficient peaks, which is accounted for by the differentiated occurrence of several levels of crystallites of a certain height or by the existence of coatings deposited on several occasions. It ranges between 0.72 (Y2) and 0.93 (Y4).

The maximum friction coefficient of the phosphate-coated specimens, the coatings of which had superior qualities revealed by the tests performed so far, is higher than the one of the reference specimen P1, treated in standard solution (0.85), and of specimens X2 and 2A (0.92) and Y4 (0.93).

The specimens which have more than two friction coefficient peaks either have needle-shaped dendritic crystallites, grown together on several levels (X2, Y4, 2B and 3A), and coarse compact and overlapping crystallites (Y1, 2A, 3B and 4B). These findings may be very well correlated with the structures revealed by microscopy.

As for the time when the friction coefficient of the metallic substrate decreases, it exceeds 200s (approx. 1300 revolutions) in the specimens with optimal protection films.

The final value of the friction coefficient of the specimens with superior properties (X2, 2A and Y4) is higher than that of the reference specimen P1 (0.58).

5.9 Analysis of Operating Behavior of the Phosphate Films

Small specimens, obtained by embossing, were used to analyze the operating behavior of phosphate films. After phosphate coating the specimens underwent a drawing procedure performed with a PHC 20 hydraulic press (Fig. 4.10) equipped with a simple cylindrical die with concentric punch.

The operating behavior of phosphate films was determined by calculating the force required to draw the phosphate coated blanks as compared to uncoated blanks.

A dynamic data collection system, which included a fully equipped master unit, force transducers and displacement transducer, was used to determine the drawing force.

Steel and phosphate-coated steel specimens, namely specimens P1, X2 and 2A, were used to compare the drawing forces.

Figure 5.45 shows the dependence of the drawing force on the punch stroke on non-coated sheet and on specimens P1, X2 and 2A.

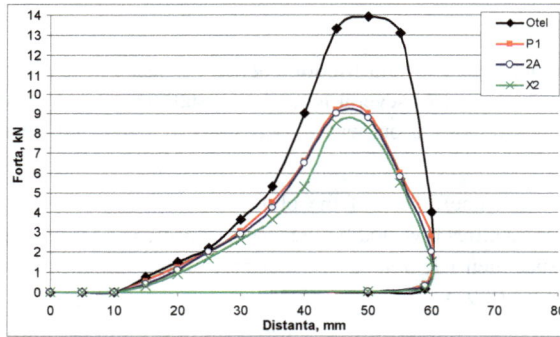

Fig. 5.45. *The variation of the drawing force applied to the specimens.*

Figure 5.46 shows the parts processed by drawing from blanks with or without phosphate coatings.

a)   b)

Figure 5.46. *Drawn parts: a) – without phosphate coating; b) – with phosphate coating.*

References

[1]   Alexander G.B., Carpenter N.F., 1976, *Method for cleaning and passivating a metal surface*, **Patent US3957529**.

[2]   Armijo, J.S., 1967, *Impurity Adsorption and Intergranular Corrosion of Austenitic Stainless in HNO_3-$K_2Cr_2O_7$ Solutions*, **Corrosion Science**, 7, 143-150. https://doi.org/10.1016/S0010-938X(67)80074-X

[3]   Amirudin, A., Thierry, D., 1996, *Corrosion mechanisms of phosphated zinc layers on steel as substrates for automotive coatings*, **Progress in Organic Coatings,** 28, 59-75. https://doi.org/10.1016/0300-9440(95)00554-4

[4]   Askienazy A., Ken V., Souchet J-C., 1980, *Phosphatation of Metallic Surfaces,* **Patent CA1071070 (A1)**;

[5]   Baciu, C., Alexandru, I., Popovici R., Baciu M., 1996, **Stiinta materialelor metalice**, Editura Didactica si Pedagogica, Bucuresti.

[6]   Banczek, E.P., Rodrigues, P.R.P., Costa, I., 2006, *Investigation on the effect of benzotriazole on the phosphating of carbon steel. Surface and Coatings Technology*, **Surface & Coatings Technology**, 201, 3701–3708. https://doi.org/10.1016/j.surfcoat.2006.09.003

[7]   Banczek, E.P., Rodrigues, P.R.P., Costa, I., 2008, *The effects of niobium and nickel on the corrosion resistance of the zinc phosphate layers*, **Surface & Coatings Technology**, 202, 2008–2014. https://doi.org/10.1016/j.surfcoat.2007.08.039

[8]   Banczek, E.P., Rodrigues, P.R.P., Costa, I., 2009, *Evaluation of porosity and discontinuities in zinc phosphate coating by means of voltametric anodic dissolution (VAD)*, **Surface and Coatings Technology**, 203, 1213–1219. https://doi.org/10.1016/j.surfcoat.2008.10.026

[9]   Bejinariu, C., Sandu, I., Predescu, A., Sandu, I.G., Baciu, C., Sandu, A.V., 2009, *New mechanisms for phosphatation of iron objects*, **Buletinul Institutului Politehnic din Iaşi**, Secţia: Ştiinţa şi Ingineria Materialelor, (ISSN 1453-1690), LV (LIX), Fasc. 1, p. 73 – 77.

[10]   Bejinariu, C., Sandu, A.V., Ionita, I., Sandu, I., Vasilache, V., Sandu, I.G, *New Procedures for Lubricant Phosphation*, **International Scientific Conference UGALMAT 2009** (Advanced Materials and Technologies), vol. II, Ed. University Press GUP (ISSN 1843-5807), Galaţi, 2009, p. 321-325;

[11]   Bejinariu, C. Sandu, I.,Vasilache, V., Sandu, I.G, Bejinariu, M.G., Sandu, A.V., Sohaciu M., Vasilache, V., 2011a, *Procedeu de fosfatare microcristalină a pieselor metalice pe bază de fier*, **Brevete RO125457-A2.**

[12]   Bejinariu, C. Sandu, I., Predescu, C., Vasilache, V., Munteanu, C., Sandu, A.V., Vasilache, V., Sandu, I.G, 2011b, *Procedeu de fosfatare cristalină lubrifianta a pieselor metalice pe bază de fier*, **Brevete RO125456-A2.**

[13]   Bentley, F.F., Smithson, L.D., Rozek, A.L., 1968, **Infrared spectra and Charactestic Frequencies**, Interscience Publishing, John Wiley & Sons, New York.

[14]   Bennett, J.M., Mattsson, L., 2008, **Introduction to Surface Roughness and Scattering,** Optical Society of America, Washington, D.C.

[15]   Berki A., Nan O., Savu C., Mitu A.M., 1998, *Lubricant, Anticorrosive, Cleaning And Passivating Composition For Metal Surfaces*, Brevet **RO113660.**

[16]   Bockris, J. O'M., Reddz, A.K.N., 1970, **Modern Electrochemistry, 2**, New York, Plenum Press.

[17]   Bojin, D., Bunea, D., Miculescu, F., Miculescu, M., 2005, **Microscopie electronică de baleiaj şi aplicaţii**, Ed. AGIR, Bucureşti.

[18]   Bosinceanu, R., Iacomi, F., Sandu, A.V., 2011, *Preparation in situ and characterization of zeolite enclosed nanoparticles*, **Journal of Optoelectronics and Advanced Materials**, 13, 9-10, 1095-1100.

[19]   Boulos, M., 2003, *Phosphate Conversion Coating Process and Composition*, **Patent US2003000418.**

[20]   Bricout J.P., Hivart P., Oudin J., Ravalard Y., 1990, *New testing procedure of zinc phosphate coatings involved in cold forging of cylindrical steel billets*, **Journal of Materials Processing Technology**, 24, 3-12. https://doi.org/10.1016/0924-0136(90)90163-O

[21]   Briggs, D., Seah, M.P. (eds), 1990, **Practical Surface Analysis**, John Wiley & Sons.

[22]   Brouwer, J.-W., Kuhm, P., 2005, *Process for Phosphate Coating Metal Surfaces Comprises Treating a Part of the Cleaning Solution and/or Rinsing Water with a Cation Exchanger and Regenerating the Exchanger after Charging with Acid*, **Patent DE10341232.**

[23]   Bunea, D., Saban, R., Toma, V., 1995, **Studiul si Ingineria Materialelor**, Editura Didactica si Pedagogica.

[24]   Callister, W.D. Jr., 1985, **Materials Science and Engineering**, John Wiley & Sons, Inc.

[25]   Creus, J., Mazille, H., Idrissi, H., 2000, *Porosity evaluation of protective coatings onto steel, through electrochemical techniques*, **Surface and Coatings Technology**, 130, 224–232. https://doi.org/10.1016/S0257-8972(99)00659-3

[26]   Chasan D.E., Ribeaud M., 2010, *Multiple Metal Corrosion Inhibitor*, **Patent US2010173808 (A1).**

[27]   Chikara O, Masanobu K, Masao T., Yoshimitsu S., Juji K., 2006, *Composite Material Having Calcium Phosphate Compound Layer And Its Production Method*, **Patent JP 2006159636(A);**

[28]   Coates, J., 2000, *Interpretation of Infrared Spectra, A Practical Approach*, **Encyclopedia of Analytical Chemistry**, R.A. Meyers (Ed.), John Wiley & Sons Ltd, Chichester.

[29]   Crow, D.R., 1994, **Principles and Applications of Electrochemistry**, Fourth Edition, Chapman & Hall, New York.

[30]   Cullity, B.D., 1978, **Elements of X-ray Diffraction**, Addison-Wesley, Reading, Mass.

[31] Czichos, H., 1978, *Tribology: a system approach to the science and technology of friction*, **Lubrication and Wear**, Elsevier Scientific Publishing Company, Amsterdam, Oxford, New York, 38–39.

[32] Dai, J., 2011, *Steel phosphorization treating fluid*, **Patent CN102021547(A).**

[33] Drake, S, 1978, **Galileo At Work**. Chicago: University of Chicago Press.

[34] Dong-Wook, K., Young-Hwa, L., 1990, *Method to Make a Phosphate Film with a Copper Coating on Steel Substrates*, **Patent KR900005843B.**

[35] Dufour, M.L., Lamouche, G., Detalle, V., Gauthier, B., Sammut, P., 2010, **Low-Coherence Interferometry, an Advanced Technique for Optical Metrology in Industry**, Industrial Materials Institute, National Research Council (Canada).

[36] Elbick, D., Prinz U., Koenigshofen A., Dahlhaus M., 2011, *Method For The Post-Treatment Of Metal Layers*, **Patent US2011272284(A1).**

[37] Etteyeb, N., Sanchez, M., Dhouibi, L., Alonso, C., Andrade, C., Triki, E., 2006, *Corrosion protection of steel reinforcement by a pretreatment in phosphate solutions: assessment of passivity by electrochemical techniques*, **Corrosion Engineering, Science and Technology**, 41, 4, 336–341. https://doi.org/10.1179/174327806X120775

[38] Fang, F., Jiang, J.H., Tan, S.Y., Ma, A.B., Jiang, J.Q., 2010, *Characteristics of a fast low-temperature zinc phosphating coating accelerated by an ECO-friendly hydroxylamine sulfate*, **Surface & Coatings Technology**, 204, 2381-2385. https://doi.org/10.1016/j.surfcoat.2010.01.005

[39] Faya, F., Linossiera, I., Langloisb, V., Harasa, D., Vallee-Rehel, K., 2005, *SEM and EDX analysis: Two powerful techniques for the study of antifouling paints*, **Progress in Organic Coatings**, 54, 3, 216–223.

[40] Flewit, P.E.J., Wild, R.K., 1994, **Physical methods for materials characterization**, IOP, Ltd., London.

[41] Flis, J., Tobiyama, Y., Mochizuki, K., Shiga, C., 1997, *Characterization of phosphate coatings on zinc, zinc-nickel and mild steel by impedance measurements in dilute sodium phosphate solutions*, **Corrosion Science**, 39, 1757–1770. https://doi.org/10.1016/S0010-938X(97)00033-4

[42] Fratesi, R., Roventi, G., 1992, *Electrodeposition of Zinc- Nickel Coatings from a Chloride Bath Containing*, **Journal of Applied Electrochemistry**, 22, 657. https://doi.org/10.1007/BF01092615

[43] Fristad, W., Saad, K., 2003, *Phosphate Conversion Coating Concentrate*, **Patent US2003209289**.

[44] Geke, J., Kuhm, P., Mayer, B., 1999, *Method of Zinc Phosphate Coating by Way of Integrated Supplementary Passivation*, **Patent PL330013**.

[45] Geru, N., Chirca, D., Bane, M., Riposan, I., Marin, M., Cosmeleata, G., Biolaru, T., 1985, **Materiale metalice – Structura, proprietati, utilizari**, Editura Tehnica, Bucuresti.

[46] Ghali, E.I., Potvin, R.J.A., 1972, *The mechanism of phosphating of steel*, **Corrosion Science**, 12, 7, 583-594. https://doi.org/10.1016/S0010-938X(72)90118-7

[47] Gheorghiu-Dobre, A., Nocivin, A., 1998, **Introducere in Stiinta Materialelor**, Editura Matrix.

[48] Girčienė, O., Gudavičiūtė, L., Juškėnas, R., Ramanauska, R., 2009, *Corrosion resistance of phosphated Zn–Ni alloy electrodeposits*, **Surface & Coatings Technology**, 203, 3072-3077. https://doi.org/10.1016/j.surfcoat.2009.03.030

[49] Gray R, Pawlik M.J, Prucnal P.J., Baldy C., 2000, *Chromium-Free Passivating Metal Substrate and Solution for Making the Same*, **Patent CZ286708.**

[50] Grünwald, E., 1995, **Tehnologii moderne de galvanizare în industria electronică şi electrotehnică**, Ed. Casa Cărţii de Ştiinţă, Cluj- Napoca.

[51] Goldstein, J., Newbury, D.E., Joy, D.C., Lyman, C.E., Echlin, P., Lifshin, E., Sawyer, L., Michael, J.R., 2003, **Scanning Electron Microscopy and X-Rays Microanalysis**, III Ed. Springer Science+Business Media, Inc.

[52] Gordin, D.M., Gloriant, T., Nemtoi, G., Chelariu, R., Aelenei, N., Guilou, A., Ansel, D., 2005, *Synthesis, structure and electrochemical behavior of a beta Ti-12Mo-5Ta alloy as new biomaterial*, **Materials Letters**, 59, 23, 2936-2941. https://doi.org/10.1016/j.matlet.2004.09.063

[53] Gosset S., Malras J-C., 1989, *Mixed phosphatation solution and process*, **Patent EP0298827** (A1);

[54]   Guan, S., Li, S., Liu, T., Ren, C., Shi, G., Wang, L., Zhang, C., 2010, *Mg-Zn based alloy phosphating solution and surface phosphating method*, **Patent CN101709465**(A).

[55]   Hamlaoui, Y., Tifouti, L., Pedraza, F., 2009, *Corrosion behaviour of molybdate–phosphate–silicate coatings on galvanized steel*, **Corrosion Science**, 51, 10, 2455–2462. https://doi.org/10.1016/j.corsci.2009.06.037

[56]   Harry, C., Cape Thomas, W., 1991, *Phosphate Coating Composition and Method of Applying a Zinc-Nickel Phosphate Coating*, **Patent KR910003722B**.

[57]   Heinz, S., Heinz, F., Josef, H., 2006, *Process and apparatus for applying a phosphate coating on workpieces*, **Patent SI0987350T**.

[58]   Ho, C.Y., Soo, K.K., 2005, *Method for inhibiting oxidation of carbon-carbon composite using oxidation-resistant phosphate coating solution*, **Patent KR20050022947**.

[59]   Iacomi, F., Calin, G., Scarlat, C., Irimia, M., Doroftei, C., Dobromir, M., Rusu, G.G., Iftimie N., Sandu, A.V., 2011, *Functional properties of nickel cobalt oxide thin films*, **THIN SOLID FILMS**, 520, 1, (2011), 651-655.

[60]   Ilca, I., Prejban, I., Petre, D., 1985, **Metalografie și proprietățile metalelor**, Litografie U.P.T.

[61]   Iliuc, I., 1980, **Tribology of Thin Layers**, Elsevier Scientific Publishing Company, Amsterdam, Oxford, New York.

[62]   Ishii, H., Nagashima, Y., 2001, *Composition and Process for Zinc Phosphate Conversion Coating*, **Patent US6231688**.

[63]   Ishizuka, K., Shindo, H., 2003, *Zinc phosphate-treated galvanized steel sheet excellent in corrosion resistance and color tone*, **Patent US6649275** B1.

[64]   Ishizuka, K., Shindo, H., Hayashi, K., 2003, *Phosphate-treated galvanized steel sheet excellent in corrosion resistance and paintability*, **Patent US6596414 B1**.

[65]   Jae-Ryung, L., 1992, *Excellent Coating Adhesive Phosphate Coating and Water Proof Adhesive Plating Steel Sheets and Process for Making*, **Patent KR920010778B**.

[66]   Jäntschi, L., 2004, **Chimie Fizică. Analize Chimice și Instrumentale**, Ed. AcademicDirect, Cluj Napoca.

[67]   Jones, D.A., 1992, **Principles and Prevention of Corrosion**, Macmillian, New York.

[68] Kasemo, B., Lausmaa, J., 1997, **Biomaterials, in: Surface Characterization: A User's Sourcebook**, (Editori: Brune D., Helborg R., Hunderi O., Whitlow H.), Scandinavian Science Publ., Weinheim.

[69] Kazuya, U., Takahiro, K., Masuru, S., 2000, *Coating Type Phosphate Treated Steel Sheet Excellent in Lubricity and Coating Material Adhesion and its Production*, **Patent JP2000064054.**

[70] Kellner, R., Mermet, J.M., Otto, M., Valcarcel M., Widmer, H.M. (editors), 2004, **Analytical Chemistry. A Modern Approach to Analytical science**, Second Edition, Wiley-VCH.

[71] Kiyokazu, I., Hidetosh, S., 2003, Zinc phosphate-treated galvanized steel sheet excellent in corrosion resistance and color tone, **Patent US6649275(B1).**

[72] Kiyokazu I., Koichi S., 2005a, *Zinc based plated strip with inorganic-organic complex treatment comprising zinc phosphate treated film and posttreated film to improve corrosion resistance and coating adhesion*, **Patent KR20050006025.**

[73] Kiyokazu I., Atsushi M., 2005b, *Zinc based plated strip with inorganic-organic complex treatment comprising zinc phosphate treated film and post treated film to improve corrosion resistance and coating adhesion*, **Patent KR20050006024.**

[74] Kolberg, T., Wietzoreck, H., Bittner, K., 2004, *Method for Applying a Phosphate Coating and Use of Metal Parts Coated in this Manner*, **Patent US2004065389.**

[75] Konishi, T., Ikeda, Y., Beppu, M., 2011, *Treatment liquid for forming protective film of steel member having nitrogen compound layer and compound layer protective film*, **Patent JP2011032512(A).**

[76] Kruger, D.H., Schneck, P., Gelderblom, H.R., 2000, *Helmut Ruska and the visualisation of viruses,* **The Lancet** 355 (9216), 1713–1717. https://doi.org/10.1016/S0140-6736(00)02250-9

[77] Li, G., Niu, L., Lian, J., Jiang, Z., 2004, *A black phosphate coating for C1008 steel*, **Surface and Coatings Technology**, 176, 215–221. https://doi.org/10.1016/S0257-8972(03)00736-9

[78] Lipson, S.G., Lipson, A., Lipson, H., 2010, **Optical Physics 4th Edition**, Cambridge University Press. https://doi.org/10.1017/CBO9780511763120

[79]   Lin, B.L., Lu, J.T., Kong, G., 2008, *Synergistic Corrosion Protection for Galvanized Steel by Phosphating and Sodium Silicate Post-Sealing*, **Surface and Coatings Technology**, 202, 1831–1838. https://doi.org/10.1016/j.surfcoat.2007.08.001

[80]   Lindert, A., Mich, T., 1983, 1. *Composition for and method of after-treatment of phosphatized metal surfaces*, **Patent US4376000.**

[81]   Liu, J., Qu, J., Zhou, X., Liu, P., Yang, Y., 2010, *Anti-corrosion paint for low heavy metal automobile parts and method for preparing same*, **Patent CN101684389(A).**

[82]   Luca, E., Barboiu, V., 1984, **Analiză structurală prin metode fizice**, vol 1 și 2, Ed. Academiei RSR, București.

[83]   Malinovschi, V., Ducu, C., 2009, **Difracția radiațiilor X pe materiale policristaline**, Ed. Universitatii din Pitești.

[84]   Marcus, Ph., 2002, **Corrosion Mechanisms in Theory and Practice**, Second Edition, Marcel Dekker Inc, New York.. https://doi.org/10.1201/9780203909188

[85]   Marinescu, A., Andonianț, Gh., Bay, E., 1984, **Tehnologii electrochimice și chimice de protecție a materialelor metalice**, Ed. Tehnică, București, 1984.

[86]   Martin, J.M., *Anti-wear mechanisms of zinc dithiophosphate: a chemical hardness approach*, **Tribology Letters**, 6, 1, 1999, 1-8.

[87]   Masada, K., Kurita, Y., 2004, *Method of Chemical Conversion Coating of Phosphate for Iron-Aluminum-and/or Zinc Substrate*, **Patent JP2004083928.**

[88]   Masahiko K., Shinichi T., 2006, *Phosphate treatment method and electro deposition coating treatment method for automobile body*, **Patent JP2006283150.**

[89]   Mateus, C., Costil, S., Bolot, R., Coddet, C., 2005, *Ceramic/fluoropolymer composite coatings by thermal spraying—a modification of surface properties*, **Surface & Coatings Technology** 191, 108–118. https://doi.org/10.1016/j.surfcoat.2004.04.084

[90]   Matsuda, H., Sakamoto, A., Horiie, N., Masuda, H., 2010, *Coating Composition Excellent In Corrosion Resistance*, **Patent JP2010001477(A).**

[91]   Michael, C.K., Gerald, C., 2001, *Zinc phosphate conversion coating and process*, **Patent KR20010086353.**

[92] Miguel, I.M., Silva, C.E.A., Peres, M.P., Voorwald, H.J.C., 2002, *Study of influence of Zinc-nickel and Cadmium Electroplated Coatings on fatique Stregth of Aeronautical Steels*, **Fatique.**

[93] Mihalcu, M., Drăgănoiu, M., 1978, **Coroziunea și combaterea în industria chimică**, Ed. Tehnica, București.

[94] Mitelea, I., Budău, V., 1987, **Studiul metalelor. Îndreptar tehnic**, Ed. Facla, Timișoara.

[95] Moore, D.M., Reynolds, R.C., 1997, *X'Ray Diffraction and the Identification and Analysis of Clay Mineral*, Oxford University Press.

[96] Munteanu, C., Ștefan, M., Baciu, C., Cimpoeșu, N., 2008, **Metode difractometrice și microscopie optică și electronică în studiul materialelor**, Ed. Tehnopress, Iași.

[97] Nakamoto, K., 1997, **Infrared and Raman Spectra of Inorganic and Coordination Compounds**, *Parts A and B*, John Wiley & Sons, New York.

[98] Narayanasamy, B., Amalraj, A.J., Selvi, J.A., Rajendran, S., 2005, *Synergistic effect of phosphate and Zn2+ on the corrosion inhibition of carbon steel*, **Bulletin of Electrochemistry**, 21, 11, 489-493.

[99] Nemtoi, G., Secula, M.S., Cretescu, I., Petrescu, S., 2007, *Studiul voltametric al dizolvarii anodice a cuprului în solutii de sulfat de cupru si acid sulfuric*, **Revista de Chimie**, 57, 10, 952-956.

[100] Nemtoi, G., Ionica, F., Lupascu, T., Cecal, L., 2010, *Voltametric characterization of the iron behaviour from steels in different electrolytic media*, **Chemistry Journal of Moldova. General, Industrial and Ecological Chemistry**, **5**, 1, 98-105.

[101] Nikdehghan, H., Amadeh, A., Honarbakhsh-Raouf, A., 2008, *Effect of substrate heat treatment on morphology and corrosion resistance of Zn–Mn phosphate coating*, **Surface Engineering**, 24, 4, 287-294. https://doi.org/10.1179/174329408X326533

[102] Ningshen, S., Kamachi Mudali, U., Amarendra, G., Baldev, R., 2009, *Corrosion assessment of nitric acid grade austenitic stainless steels*, **Corrosion Science**, 51, 322-329. https://doi.org/10.1016/j.corsci.2008.09.038

[103] Noriaki K., Tomohiro O., 2007, *Method for manufacturing material provided with phosphate coating film*, **Patent JP2007146221.**

[104] Ogle, K., Tomandl, A., Meddahi, N., Wolpers, M., 2004, *The alkaline stability of phosphate coatings I: ICP atomic emission spectroelectrochemistry*, **Corrosion Science**, 46, 2004, 979–995. https://doi.org/10.1016/S0010-938X(03)00182-3

[105] Okada, K., Kawamura, Y., Takahashi., M., Hiroaka, O., Yoshioka A., Futaba, K., Matsumoto, M., Ueda, N., 2011, *Coated metallic material, solution for chemical conversion treatment for manufacturing the coated metallic material, and casing formed by using the coated metallic material*, **Patent JP2011038139 (A).**

[106] Oniciu, L., Grüwald, E., 1980, **Galvanotehnica**, Ed. Ştiinţifică şi Enciclopedică, Bucureşti, 1980.

[107] Oniciu, L., Constantinescu E., 1982, **Electrochimie şi coroziune**, Ed. Didactică şi Pedagogică, Bucureşti.

[108] Pareek S.R., Rajsharad C., 2010, *Film Coating Compositions Based On Tribasic Calcium Phosphate*, **Patent MX 2010002026(A);**

[109] Pascariu, P., Tanase, S.I., Pinzaru, D., Sandu, A.V., Georgescu, V., 2012, *Preparation and magnetic properties of electrodeposited [Co/Zn] multilayer films*, **Materials Chemistry and Physics**, 131, 3, pp. 561–568. https://doi.org/10.1016/j.matchemphys.2011.10.021

[110] Paukson, A., 2011, *Iron Phosphate Composition And Method For Preparing And Use Thereof*, **Patent WO2011086526(A1).**

[111] Petraco, N., Kubic, T., 2003, **Microscopy for criminalists, chemists and conservators**, CRC Press, New York.

[112] Pinzaru, D., Tanase, S.I., Pascariu, P., Sandu, A.V., Nica, V., Georgescu, V., 2011, *Magnetic properties and giant magnetoresistance effect in [Fe/Pt]n granular multilayers*, **Optoelectronics and Advanced Materials – Rapid Communication**, 5, 3-4, 235-241.

[113] Ploaie, P.G., Petre, Z., 1979, **Introducere în Microscopia Electronică**, Ed. Academiei RSR, Bucureşti.

[114] Poiană, M., Dobromir, M., Sandu, A.V., Georgescu, V., 2012a, *Investigation of Structural, Magnetic and Magnetotransport Properties of Electrodeposited Co–TiO2 Nanocomposite Films*, **Journal of Superconductivity and Novel Magnetism**, 25. https://doi.org/10.1007/s10948-012-1612-3

[115] Poiana, M., Vlad, L., Pascariu, P., Sandu, A.V., Nica, V., Georgescu, V., 2012b, *Effects of current density on morphology and magnetic properties of Co-TiO₂ electrodeposited nanocomposite films*, **Optoelectronics and Advanced Materials, Rapid Communications,** 6 (3-4), pp. 434-440.

[116] Preston, D.W., Dietz, E.R., 1991, **The Art of Experimental Physics**, Wiley.

[117] Primer, A., 1996, **Fundamentals of modern UV-visible spectroscopy**, Hewlett-Packard.

[118] Qiu, G., Yin, H., Liu, F., Feng, X., Tan, W., Liu, M., Chen, X., 2010, *Method for preparing manganese phosphate material*, **Patent CN101891177(A).**

[119] Rausch, W., 1990, **The Phosphating of Metals**, Finishing Publications Ltd., London.

[120] Ruska, E., 1986, **Autobiografia lui Ernst**, Nobel Foundation, http://nobelprize.org/nobel_prizes/physics/laureates/1986/ruska-autobio.html;

[121] Salzer, R., Siesler H.W. (editori), 2009, **Infrared and Raman Spectroscopic Imaging**, Wiley-VCH.

[122] Sànchez, M., Gegori, J., Alonso, M.C., Garcia-Jareno, J.J., Vicente, F., 2006, *Anodic growth of passive layers on steel rebars in an alkaline medium simulating the concrete pores,* **Electrochemica Acta, 52**, 47-53. https://doi.org/10.1016/j.electacta.2006.03.071

[123] Sànchez M., Gegori J, Alonso M.C., Garcia-Jareno J.J., Takenouti H., Vicente F., 2007, **Electrochemica Acta**, 52, 27, 7634-7641. https://doi.org/10.1016/j.electacta.2007.02.012

[124] Sandu, A.V., Bejinariu, C., 2010a, *New Phosphated Layers on Iron Support With Lubricant Properties*, **Buletinul Institutului Politehnic din Iaşi, Secţia Ştiinţa şi Ingineria Materialelor**, Tomul LVI (LX), Fasc. 4, (2010), p. 97-102;

[125] Sandu, A.V., Bejinariu, C., 2010b, *Obtaining and Characterization of superficial phosphated layers on iron support*, **Buletinul Institutului Politehnic din Iaşi, Secţia Ştiinţa şi Ingineria Materialelor**, Tomul LVI (LX), Fasc. 2, 113-116.

[126] Sandu, A.V., Bejinariu, C., 2011, Predescu, A., Sandu, I.G, Baciu C., Sandu, I., 2011, *New mechanisms for chemical phosphatation of iron objects*, **Recent Patent on Corrosion Science**, (ISSN 1877-6108), 1, 1, 33-37.

[127] Sandu, A.V., Coddet, C., Bejinariu, C., 2012a, *A Comparative Study on Surface Structure of Thin Zinc Phosphates Layers Obtained Using Different Deposition*

Procedures on Steel, **Revista de Chimie**, 63, 4, pp. 401-406.

[128] Sandu, A.V., Coddet, C., Bejinariu, C., 2012b, *Study on the chemical deposition on steel of zinc phosphate with other metallic cations and hexamethilen tetramine. I. Obtaining and Structural and Chemical Characterization,* **Journal of Optoelectronics and Advanced Materials**, 14, 7-8, 699-703.

[129] Sandu, A.V., Ciomaga, A., G. Nemtoi, Bejinariu, C., Sandu, I., 2012c, *Study on the chemical deposition on steel of zinc phosphate with other metallic cations and hexamethilen tetramine. II. Evaluation of corrosion resistance,* **Journal of Optoelectronics and Advanced Materials**, 14, 7-8, 704-708.

[130] Sandu, A.V., Ciomaga, A., G. Nemtoi, Bejinariu, C., Sandu, I., 2012d, *SEM-EDX and microFTIR studies on evaluation of protection capacity of some thin phosphate layers,* **Microscopy, Research and Technique**, 75, online first. https://doi.org/10.1002/jemt.22120

[131] Sandu, A.V., Bejinariu, C., Sandu, I.G., Baciu, C., 2012e, *Morphologic characterization of some thin zinc phosphate layers*, **Buletinul Institutului Politehnic din Iaşi, Secţia Ştiinţa şi Ingineria Materialelor**, Tomul LVIII (LXII), Fasc. 1, 9-14.

[132] Sandu, A.V., Bejinariu, C., Sandu, I.G., Baciu, C., 2012f, *Tribological study on some thin zinc phosphate layers*, **Buletinul Institutului Politehnic din Iaşi, Secţia Ştiinţa şi Ingineria Materialelor**, Tomul LVIII (LXII), Fasc. 2, 51-55.

[133] Sandu, A.V., Bejinariu, C., Sandu, I.G., Vizureanu, P., Sandu, I., Vasilache, V., 2012g, *Procedeu de protecţie anticorozivă a pieselor din fier prin fosfatare in sistem apos,* **Dosar OSIM a 2012 00144/** 07.03.2012.

[134] Sandu, A.V., Bejinariu, C., Sandu, I.G., Ionita, I., Sandu, I., Baciu, C., Vasilache, V., 2012h, *Procedeu de fosfatare anticoroziva a pieselor metalice din fier,* **Dosar OSIM a 2012 00146/** 07.03.2012.

[135] Sandu, I., Dima, A., Sandu, I.G, 2002, **Restaurarea şi conservarea obiectelor metalice**, Ed. Corson, Iaşi.

[136] Sandu, I.G., Dima, A., Sandu, I., Roibu, L., Sandu, A.V., Roibu, L.O., 2009, *Procedeu de pasivare prin fosfatare cristalină a pieselor din fier,* **Brevet RO122303 - B1.**

[137] Satoh, N., 1987, *Effects of heavy metal additions and crystal modification on the zinc phosphating of electrogalvanized steel sheet,* **Surface and Coatings Technology**, 30, 171–181. https://doi.org/10.1016/0257-8972(87)90141-1

[138] Satoh, N., Minami, T., 1988, *Relationship between the formation of zinc phosphate crystals and their electrochemical properties,* **Surface and Coatings Technology**, 34, 3, 331–343. https://doi.org/10.1016/0257-8972(88)90123-5

[139] Sato, N., Minami, T., Kono, H., 1989, *Analysis of metallic components in zinc phosphate films using electron spin resonance and X-ray fluorescence,* **Surface and Coatings Technology**, 37, 23–30. https://doi.org/10.1016/0257-8972(89)90118-7

[140] Schlesinger, M., 2000, **Electrodeposition of Alloys, Modern Electroplating**, Fourth Edition, John Wiley and Sons, Inc. New-York.

[141] Schoenherr M., Wawrzyniak J.Z., Wiedemann E., 2009, *Zirconium phosphating of metal components, in particular iron,* **Patent WO2009068523 (A1).**

[142] **Scott, D.A., 1991,** Metallography and microstructure of ancient and historic metals, **The Getty Conservation Institute.**

[143] Seidel R., Brands K.D., Gottwald, K.H., 1995, *Process for the passivating post-treatment of phosphatized metal surfaces,* **Patent US5391240.**

[144] Sheng, M., Wang, Y., Zhong, Q., Wu, H., Zhou, Q., Lin, H., 2011, *The effects of nano-SiO2 additive on the zinc phosphating of carbon steel,* **Surface & Coatings Technology**, 205, 3455-3460. https://doi.org/10.1016/j.surfcoat.2010.12.011

[145] Sigeki, M. 1992, *Method of Forming a Chemical Phosphate Coating on the Surface of Steel,* **Patent KR890004789B.**

[146] Soares, M.E., Souza, C.A.C., Kuri, S.E., 2006, *Corrosion resitence of Zn-Ni electrodeposited alloy obtained with a controlled electrolyte flow and gelatin additive,* **Science Direct,** 201, 6, 2953-2959.

[147] Song-Gu, C., 1988, *Phosphate Coating Solution and Using Method for Normal Temperature,* **Patent KR880001108B.**

[148] Stern, M., Geary, A.L., 1957, *The Mechanism of Passivating-Type Inhibitors,* **Journal Electrochemistry Society**, 105, 638-647. https://doi.org/10.1149/1.2428683

[149] Stout, K J., Blunt, L., 2000, **Three-Dimensional Surface Topograhy** (2nd ed.), Penton Press.

[150] Suciu, V., Suciu, M.V., 2008, **Studiul Materialelor**, Ed. Fair Partner.

[151] Sugama, T., Kukacka, L. E., Carciello, N., Warren, J. B., 1988, *Aspects of the adhesion and corrosion resistance of polyelectrolyte-chemisorbed zinc phosphate*

conversion coatings, **Journal of Materials Science**, 23, 1, 101-110.
https://doi.org/10.1007/BF01174040

[152] Sugama, T., Broyer, R., 1992, *Advanced Poly(acrylic) acid - Modified Zinc Phosphate Conversion Coatings: Use of Cobalt and Nickel Cations,* **Surface & Coatings Technology**, 50, 89-95. https://doi.org/10.1016/0257-8972(92)90048-F

[153] Tamotsu, S., Teturo, K., Minoru, I, 1999, *Process for Phosphating Metal Surface to form a Zinc Phosphate,* **Patent KR183023B**.

[154] Tanase, S.I., Tanase, D., Sandu, A.V., Georgescu, V., 2012, *Magnetic Field Effects on Surface Morphology and Magnetic Properties of Co–Ni–N Thin Films Prepared by Electrodeposition,* **Journal of Superconductivity and Novel Magnetism**, 25. https://doi.org/10.1007/s10948-012-1562-9

[155] Tanase, S.I., Pinzaru (Tanase), D., Pascariu, P., Dobromir, M., Sandu, A.V., Georgescu., V., 2011, *Effect of nitrogen addition on the morphology, magnetic and magnetoresistance properties of electrodeposited Co, Ni and Co–Ni granular thin films onto aluminum substrates,* **Materials Chemistry and Physics**, 130, 327– 333. https://doi.org/10.1016/j.matchemphys.2011.06.055

[156] Tanase, S.I., Tanase, D., Pascariu, P., Vlad, L., Sandu, A.V., Georgescu, V., 2010, *Tunneling magnetoresistance in Co–Ni–N/Al granular thin films*, **Materials Science and Engineering B**, 167, 119–123. https://doi.org/10.1016/j.mseb.2010.01.061

[157] Taranu, I., Fagadar, C.G., Goanta, I., Radoi, I., 1997, *Soluție de fosfatare,* **Patent RO113665**;

[158] Townsend, H.E., **NACE Annual Corrosion Conference**, Cincinnati, OH, 1991. Paper No. 416.

[159] Trusov, V.I., Kiselev, V.L., 2004, *Composition for Phosphatation of Metallic Surfaces,* **Patent RU2241069** (C2);

[160] Tsai, C.Y., Liu, J.S., Chen, P.L., Lin, C.S., 2010a, *Effect of Mg^{2+} on the Microstructure and Corrosion Resistance of the Phosphate Conversion Coating on Hot-dip Galvanized Sheet Steel,* **Corrosion Science**, 52, 3907-3916. https://doi.org/10.1016/j.corsci.2010.08.007

[161] Tsai, C.Y., Liu, J.S., Chen, P.L., Lin, C.S., 2010b, *A Two-step Roll Coating Phosphate/Molybdate Passivation Treatment for Hot-dip Galvanized Steel Sheet*, **Corrosion Science**, 52, 3385-3393. https://doi.org/10.1016/j.corsci.2010.06.020

[162] Tsai, C.Y., Liu, J.S., Chen, P.L., Lin, C.S., 2011, *A Roll Coating Tungstate Passivation Treatment for Hot-dip Galvanized Sheet Steel*, **Surface & Coatings Technology**, 205, 5124-5129. https://doi.org/10.1016/j.surfcoat.2011.05.022

[163] Tzeng, C.-J., 2004, *Method of Phosphate Surface Treatment on Metal Surface for Subsequent Coating*, **Patent TW593752B.**

[164] Urata, K., Kubota, T., Sagiyama, M., 2000, *Coating Type Phosphate Steel Treated Steel Sheet in Lubricity and Coating Material Adhesion, and its Production*, **Patent JP2000064054.**

[165] Urata, K., Miyoshi, T., Kubota, T., Yamashita, M., 2002, *Steel Sheet Coated with Composite Phosphate Film Superior in Corrosion Resistance, Lubricity and Coating Material Adhesiveness*, **Patent JP2002012983.**

[166] Varela, L.F., 2006, *Method for Applying a Phosphate Coating on a Steel or Iron Part and Corresponding Steel or Iron Part*, **Patent EP1660699.**

[167] Varentsova N.V., Chumaevskij V.A., 1998, *Solution for Phosphatation of Metal Surface*, **Patent RU2111282**(C1).

[168] Vlad, L., Sandu, A.V., Georgescu, V., 2012, *The Effects of the Thermal Treatment on the Structural and Magnetic Properties of Zn–Co Alloys Prepared by Electrochemical Deposition*, **Journal of Superconductivity and Novel Magnetism**, 25, 469-474. https://doi.org/10.1007/s10948-011-1301-7

[169] Walls, J.M. (ed), 1989, **Methods of Surface Analysis**, Cambridge Univ. Press.

[170] Warren, B.E., 1990, **X-ray Diffraction**, Dover Publications, 1990.

[171] Wen, N.T., Lin, C.S., Bai, C.Y., Ger, M.D., 2008, *Structures and Characteristics of Cr(III)-based Conversion Coatings on Electrogalvanized Steels*, **Surface and Coatings Technology**, 203, 317-323. https://doi.org/10.1016/j.surfcoat.2008.09.006

[172] Xu, Y., Yang, Y., Li, D., Ji, Y., Zhao, Z., Chen, L., Chen, X., Huang, J., 2012, *Non-Chromic Insulating Coating For Non-Oriented Silicon Steel*, **Patent WO2012041052 (A1).**

[173] Yoshio, N., Takashi, K., Masanori, K., 1988, *Preparing Metals for Cold Forming: Forming Zinc Containing Phosphate and Lubricating Coating*, **Patent NZ215988.**

[174] Zhang, G., Liao, H., Li, H., Mateus, C., Bordes, J.M., Coddet, C., 2006, On dry *sliding friction and wear behaviour of PEEK and PEEK/SiC-composite coatings*, **Wear** 260, 594–600. https://doi.org/10.1016/j.wear.2005.03.017

[175] Zarei, A., Afshar, A., TMS 2009: **Materials Processing and Properties**, 2009, 615-623.

[176] Zimmermann, D., Munoz, A.G., Schultze, J.W., 2003, *Microscopic local elements in the phosphating process*, **Electrochimica Acta**, 48, 3267–3277. https://doi.org/10.1016/S0013-4686(03)00385-2

[177] Zimmermann, D., Munoz, A.G., Schultze, J.W., 2005, *Formation of Zn-Ni alloys in the phosphating of Zn layers*, **Surface and Coatings Technology**, 197, 260–269. https://doi.org/10.1016/j.surfcoat.2004.07.129

[178] Zhang, G., Yub, H, Zhang, C., Liao, H., Coddet, C., 2008, *Temperature dependence of the tribological mechanisms of amorphous PEEK (polyetheretherketone) under dry sliding conditions,* **Acta Materialia** 56, 2182–2190. https://doi.org/10.1016/j.actamat.2008.01.018

[179] Zhang Z., 2009, *Microcrystalline phosphating solution for brake block hardware,* **Patent CN101348906 (A).**

[180] **CSM Tribometers**, http://www.csm-instruments.com

[181] http://www.tescan.com

[182] http://web.mit.edu/Invent/iow/hillier.html - Biografia lui Hillier;

[183] http://www.zeiss.com/c12567be0045acf1/Contents-Frame/15c864640a87573ac12577040034a0f0

[184] http://www.bruker-axs.com/d8_focus.html

Keyword Index

About the Authors

Andrei Victor SANDU

Associate Professor PhD.Eng. (Lecturer)

Gheorghe Asachi Technical University of Iasi

President of Romanian Inventors Forum

sav@tuiasi.ro, www.afir.org.ro/sav

Researcher and associate professor at Gheorghe Asachi Technical University of Iasi, Faculty of Materials Science and Engineering, Dr. Sandu has expertise in the field of Materials Science, mainly on advanced analysis techniques. He has started young his "scientific life" with a first publication at 18 years old. Now he has over 250 publications, 160 of them indexed by Web of Science (Thomson Reuters) and over 30 patents. Regarding international recognition, the Hirsh index is 18 (over 1000 citations), being a visiting Professor at Universiti Malaysia Perlis. On the innovative side, he has received over 100 medals on Inventions Exhibitions and Contests and various important orders. He is the President of the Romanian Inventors Forum, member of WIIPA – World Invention Intellectual Property Associations and full member for Romania at IFIA – International Federation of Inventors' Associations. He is publishing editor of International Journal of Conservation Science (Web of Science and Scopus Indexed) and European Journal of Materials Science and Engineering and reviewer for many valuable journals.

Costica BEJINARIU

Professor Ph.D. Eng.

Vicedean of Faculty of Materials Science and Engineering,

"Gheorghe Asachi" Technical University of Iasi

costica.bejinariu@tuiasi.ro

Professor and researcher at "Gheorghe Asachi" Technical University of Iasi, with more than 30 years of experience. PhD Coordinator since 2009, with 2 granted PhD students and 7 undergoing PhD student. Field of experience is Materials Engineering with 15 published books over 170 published articles with more than 200 citations – H index of 10. Worked on more than 45 research grants, on 4 being director and another 3 institution responsible. He has 12 patents and many awards received for them. He is a member of various academic societies and also reviewers for many scientific journals and conferences.

Ioan Gabriel SANDU

Associate Professor PhD.Eng.(Lecturer)
„Gheorghe Asachi" Technical University of Iasi
gisandu@yahoo.com

Ioan Gabriel SANDU, is Associate Professor (Lecturer) at „Gheorghe Asachi" Technical University of Iasi, Faculty of Materials Science and Engineering and Associate Researcher for Center of Excellence Geopolymer & Green Technology Faculty of Engineering Technology Universiti Malaysia Perlis (UniMAP). His professional, teaching and researching experience, which started in 2006, allowed him to improve her competence in surface engineering. He has over 150 publications with H Index of 17 (Web of Science) and various awards received for patents.

Mohd Mustafa Al Bakri ABDULLAH,

Associate Professor PhD.
Universiti Malaysia Perlis, Faculty Engineering Technology (FETech)

Prof. Mustafa is a Pioneer on Engineering, mainly Geopolymer science. He is also Manager of Centre of Excellence Geopoymer & Green Technology (CEGeoGTech), Vice President of Malaysian Geopolymer Society (MyGeopolymer) and Vice President of World Invention Intellectual Property Association (WIIPA). He has published over 300 articles, having H-index 18. He is coordinator of several international research projects.

www.ingramcontent.com/pod-product-compliance
Lightning Source LLC
Chambersburg PA
CBHW071651210326
41597CB00017B/2175